How a Brilliant Idea Can Transform the Field of Particle Detectors

How a Brilliant Idea Can Transform the Field of Particle Detectors

Homage to Georges Charpak

By

Giacinto de Cataldo and Vladimir Peskov

Cambridge
Scholars
Publishing

How a Brilliant Idea Can Transform the Field of Particle Detectors

By Giacinto de Cataldo and Vladimir Peskov

This book first published 2026

Cambridge Scholars Publishing

Lady Stephenson Library, Newcastle upon Tyne, NE6 2PA, UK

British Library Cataloguing in Publication Data
A catalogue record for this book is available from the British Library

ISBN: 978-1-0364-6674-9
ISBN (Ebook): 978-1-0364-6675-6

To my brotherly friend and university colleague, Piero Novielli and to my dearest Anna, Silvia, Gianluca and Francesco.

*—**GDC***

To my wife, Tatiana Peskova, whose unwavering support has guided me throughout my scientific journey.

—V. P.

Contents

Preface xii

About the Authors xiii

Introduction xv

Chapter 1 1
A Bit of Detector History
Abstract 1
1.1. Ionization chambers 3
1.2. Geiger-Muller counters 4
1.3. Cloud chambers, bubble chambers and spark counters 6
1.4. Multiwire proportional chambers and their descendants 14
1.5. Other imaging detectors 19
Concluding Remarks 23

Chapter 2 24
From Precursors to Innovation: Charpak's Legacy and the Fundamentals of Avalanche Gaseous Detectors
Abstract 24
2.1. Townsend avalanche 24
2.2. Geiger and proportional counter 28
2.3. Operational features of single-wire Counters 40
2.4. Parallel-plate avalanche chamber 54
Concluding Remarks 62

Chapter 3 63
The MWPC and Mesh PPAC
Abstract 63
3.1. MWPC, principle of operation, and main characteristics 63
3.2. Mesh PPAC 75
3.3. Concluding remarks 75

Chapter 4 77
MWPC Conditioning
Abstract 77
4.1. Initial problems with MWPC 77
4.2. A multistep detector 79
4.3. Improved designs of MWPC 85
Concluding remarks 89
Optional Advanced Reading 90

Chapter 5 96
MWPC and PPAC Descendants
Abstract 96
Introduction 96
5.1. Drift chamber 99
5.2. Time projection chamber 102
5.3. Photosensitive MWPC 107
5.4. Optical imaging chambers 111
Concluding remarks 114
Additional reading 115

Chapter 6 116
The MWPC for Particle Identification, Calorimeters and Muon Stations
Abstract 116
Introduction 116
6.1. MWPC for TRD detectors 118
6.2. MWPCs for RICH detectors 119
6.2.1. TMAE based RICH 121
6.2.2. CsI-based RICH 123
6.3. MWPC for calorimeters 125
6.4. MWPCs in muon detectors 127
Concluding remarks 128
For the Curious: The Transition Radiation and Cherenkov radiation 128

Chapter 7 134
MWPC Applications in High-Energy Physics Experiments
Abstract 134
Introduction 134
7.1. Review of MWPC applications in high-energy physics experiments (1975-1989) 135
7.2. Experiments in the LEP era (1989-2000) 144

7.3. Some other experiments 148
7.4. LHC era (2008-present time) 149
7.5. Concluding remarks 154
7.6. Level Up: MWPCs in the ALICE experiment 154
7.6.1 The MWPC for the ALICE HMPID 155
7.6.2. The MWPC for the ALICE TPC 161
7.6.3. The MWPC for the ALICE Transition Radiation Detectors 166
Concluding remarks 169

Chapter 8 170
Probing the Cosmos: Applications of MWPCs in Astrophysics
Abstract 170
Introduction 170
8.1. The MWPCs in the cosmic rays study 171
8.1.1 Ground based experiments 171
8.1.2 Balloon-borne Experiments 174
8.1.3. Satellite experiments 179
Concluding remarks 181

Chapter 9 182
Optical Chambers: Potential Applications and Future Directions
9.1. Optical chamber for neutron imaging 185
9.2. Optical TPC 186
9.3. Prototypes of High-Pressure TPC for the DUNE Experiment 189
9.4. ARIADNE project 191
Concluding Remarks 193

Chapter 10 194
Physics Discoveries and Charpak's Inventions: The Interplay between Particle Detectors and Theory
10.1. Discovery of the J/ψ particle containing the new charm quark 194
10.2. Discovery of W and Z^0 196
10.3. For the curious: What antimatter reveals in the primary cosmic rays 202
Concluding remarks 209

Chapter 11 211
MWPCs Application beyond HEP and Astrophysics
Abstract 211
Introduction 211
11.1. Early applications in medicine, biology, and crystallography 212

11.2. Applications in Security devices and in muon muonography and tomography ... 213
11.3. Plasma diagnostics ... 215
11.4. Some ongoing R&D oriented on new potential applications ... 218
11.4.1. MWPCs in PET ... 218
11.4.2. Neutron detectors ... 218
Concluding remarks ... 219

Chapter 12 ... 220
Charpak's Final Contributions to Detector Development
Abstract ... 220
12.1. Invention of MICROMEGAS ... 220
12.1.1 Introduction: The history of development ... 220
12.1.2. The main characteristics of MICROMEGAS ... 224
12.1.3. MICROMEGAS with resistive electrodes ... 229
12.1.4. MICROMEGAS applications ... 231
12.1.5. Concluding Remarks ... 231
12.2. George's final research projects ... 232
12.2.1. Low-cost gaseous detectors for massive online monitoring of Rn in seismic active regions ... 233
12.2.2. Sensors for detecting hazardous and flammable gases ... 238
12.2.3. Design, tests and optimization of devices for early forest fire detection ... 240
12.2.4. Advanced Manufacturing of High-Efficiency Solar-Blind Photocathodes ... 243
12.2.5. Concluding Remarks ... 245

Chapter 13 ... 247
Electronic Components and AI: Crucial Elements of MWPCs
Abstract ... 247
13.1 Electronic block diagram for the TRD of TS93 experiment ... 249
13.2. Electronic Block diagram for the ALICE HMPID at CERN ... 250
13.3. Electronic Block diagram for the ALICE TPC (Run 2) ... 252
13.4. The Role of Artificial Intelligence in analysing Physics Data ... 254
Concluding remarks ... 258

Final Thoughts ... 260

Appendix ... 263
For Curious Readers: Further Exploration

Biography of George Charpak 270

Some Fun Stories About Charpak 279

References 306
    Further Readings 324

Acknowledgments 326

List of Abbreviations 327

PREFACE

Georges Charpak was awarded the Nobel Prize in Physics in October 1992 for his invention of the Multiwire Proportional Chamber (MWPC), a groundbreaking detector that has played a pivotal role in numerous major advances in physics. While several technical publications have thoroughly detailed the underlying physical principles of the MWPC, this book takes a different approach.

Here, we explore the innovative ways in which the MWPC has been employed in major physics discoveries, as well as its applications beyond the realm of particle physics enabled by its unique capabilities and versatility.

Over the course of our careers, both authors have acquired extensive experience in the design and use of gaseous detectors, working on collider experiments, spectrometers, and underground instruments for cosmic ray research, along with various interdisciplinary applications. One of us had the privilege of knowing and collaborating with Georges Charpak personally, a connection that inspired us to honour his many brilliant contributions to particle detector technology.

This book is aimed at a broad audience, from students to young researchers and engineers. It combines accessible sections introducing fundamental concepts with more advanced material requiring a solid grasp of instrumentation and particle physics. We hope it inspires young researchers to embrace creativity and curiosity in their work, contributing to the ongoing advancement of science and technology.

ABOUT THE AUTHORS

Giacinto De Cataldo

G. De Cataldo is an Italian physicist of the *Istituto Nazionale di Fisica Nucleare (INFN),* leading two international projects within the ALICE experiment at CERN. He also lectures in Particle Physics (PHY3232) at the University of Malta. From 1987 to 2000, he was involved in research on cosmic ray composition and participated in R&D activities for a large-area Transition Radiation Detector (TRD) designed for installation in underground laboratories for cosmic ray studies. Beginning in 1995, he participated in two missions in collaboration with the University of New Mexico (USA) and NASA, aimed at searching for primordial antimatter in primary cosmic rays using spectrometers flown on helium-filled stratospheric balloons. These spectrometers featured compact TRDs based on Multi-Wire Proportional Chambers and straw tubes. Within the same collaboration, he led the development of a position-sensitive Time-of-Flight system for a satellite-based spectrometer, employing neural network signal processing. In 1997, he worked on the design of a large-area calorimeter integrating a TRD for muon detection, targeting the study of neutrino oscillations over a 700 km baseline between CERN and the Gran Sasso Laboratory, Italy. Since 2000, he has been a member of the ALICE collaboration at CERN, focusing on the study of the Quark–Gluon Plasma (QGP) via ultra-relativistic heavy-ion collisions at the Large Hadron Collider (LHC). From 2002 to 2004, he served as a paid scientific associate at CERN. In 2010, he was appointed Project Leader of the ALICE–LHC Interface project, and from 2012 to 2023, he led the High Momentum Particle Identification detector project within ALICE. G. De Cataldo is the author or co-author of more than 500 scientific publications, spanning research on cosmic rays, neutrino oscillations, the QGP, and the development of gaseous detectors.

Vladimir Peskov

Vladimir Peskov is a professor and visiting scientist at CERN, currently affiliated with the University of Bari in Italy. From 1971 to 1985, he worked at the Physics Laboratory of the Russian Academy of Sciences in Moscow, led by Nobel laureate P.L. Kapitza, where he conducted research on ultrahigh-frequency plasma phenomena. During this period, he developed innovative plasma diagnostics methods based on MWPC, which led to the discovery of new phenomena in plasma physics.

From 1986 to 1992, Peskov was a scientific associate at CERN, working in the Charpak group. His primary achievement during this period was the development of gaseous detectors combined with CsI and other solid photocathodes. These advancements led to a new generation of Ring Imaging Detectors, later used in ALICE and other high-energy physics experiments.

Between 1992 and 2004, Peskov worked at various research centres, including Fermilab, NASA, and the Royal Institute of Technology in Stockholm. From 2004 to 2010, he closely collaborated with G. Charpak on the development of supersensitive detectors for sparks, flames, and dangerous gases, aimed at enhancing environmental safety. From 2004 to 2006, he was a member of the nTOF experiment at CERN, and since 2006, he has been involved with the ALICE experiment at the LHC.

Since 2008, he participated in the RD51 project, where he played a key role in developing spark-protective micropattern detectors. Peskov is also the author and co-author of more than 300 publications and four books on gaseous detectors of ionizing radiation.

INTRODUCTION

Why was that title chosen for the book?

We aim to illustrate to readers (particularly young ones aspiring to join the realm of physics science) how a brilliant and seemingly simple concept transformed significantly the area of particle physics detectors.

It is important to highlight that, generally, both science and technology do not progress steadily and uniformly over time; instead, they advance through abrupt breakthroughs made by the input of geniuses (exceptionally talented individuals). Numerous general examples exist; let’s highlight two randomly selected ones: Edison’s major inventions (the light bulb, movies, phonograph...) and the groundbreaking methods brought forth by W. Roentgen (radiology, medical research utilizing X-rays).

The examples can certainly be extended: Marconi, Bell, etc. However, what led us to select G. Charpak? What is his rank among other inventors? Are his contributions comparable to those of other renowned inventors?

Making a direct comparison is not an easy task since the inventions mentioned pertain to various disciplines. Edison and Röntgen's inventions significantly enhanced the quality of our everyday lives, while Marconi's radio, linking everyone globally, sparked the first instances of 'globalization.' In comparison, Charpak’s research focuses on experiments within fundamental physics, examining the essential components of matter and the interactions between them. In this intriguing discipline, a unique role is held by the so-called “imaging detectors,” which enable the visualization of particle trajectory in three-dimensional space. Through the study of particle kinematics, it is possible to explore the features of the fundamental interactions and to infer fundamental physical laws.

Physics is filled with innovative concepts that have become essential for advancing understanding in that area. In physics particle detectors, the Multiwire Proportional Chamber (MWPC) is among those concepts.

George Charpak invented it in 1968 for tracking, leading to a revolution in detection methods. It is a gas detector utilizing parallel elongated wires, several tens of microns in diameter, positioned at a spacing of several millimetres, with a length of nearly one meter. In this grid, situated between two cathodic planes, every wire functions as an individual proportional counter. This design allows for the construction of detectors with active areas ranging from tens of square centimetres to several square meters.

During the initial prototype phase, integrated circuits containing hundreds of transistors, significantly smaller than bulky electronic tubes, emerged following the USA's aerospace missions. Right away, researchers at the European Organization for Nuclear Research (CERN) began outfitting every wire with electronic circuits that incorporated all necessary components for signal processing and data storage on computers. Custom electronic read-out systems with thousands of channels were subsequently designed and built specifically for MWPCs.

The MWPC enabled, for the first time, the electronic capture of charged particles and energetic photon events at a high rate.

Shortly, two MWPCs configured with perpendicular wires, or a single detector with resistive wires and left-right signal sharing, were employed as trackers in fixed-target experiments. They offered two-dimensional (2-D) readouts with high positional accuracy, surpassing the wire pitch.

An additional clever concept emerged from G. Charpak. Indeed, he investigated the "drift time" of groups of electrons, and in 1971 this concept was utilized for the drift chamber and subsequently for the Time Projection Chamber.

The drift chamber, featuring a simpler design compared to two perpendicular MWPCs, could deliver the two coordinates of the event with great accuracy: the first from the affected wire (like the MWPC), and the second, at right angles, from the drift time of the electron cluster to reach the wire.

Subsequently, MWPCs, drift chambers, and magnetic dipoles were combined in spectrometers to conduct experiments at fixed target accelerators or colliders. Significant discoveries occurred, such as the charm quark at the AGS proton accelerator of Brookhaven in 1974 (J.J. Aubert, 1974, 1404-1406) and, around the same period, the affirmation at the SPEAR e+e- collider at Stanford Linear Accelerator Center [Augustin , 1974, 1406-1408].

The time projection chamber, employing MWPCs as readout detectors positioned at the endcaps of a cylindrical drift region filled with noble gas, successfully generates three-dimensional representations of charged particle tracks. By introducing a magnetic field into the gas volume and measuring the ionization density along the particle trajectory, this detector, alongside its strong tracking capabilities, was inherently equipped with efficient particle identification functionality.

In the 1980s, detectors like drift chambers and time projection chambers enabled physicists to achieve significant breakthroughs in high-energy physics experiments.

In 1983, C. Rubbia and his team in the UA1 and UA2 experiments at the CERN Super Proton Synchrotron collider identified the W and Z° bosons, which are the heavy mediators of the weak force.

In 1989, G. Charpak's next innovative concept was to investigate the capabilities of the MWPC utilizing CsI-activated photocathodes for detecting ultraviolet photons.

In the 1990s, the MWPC featuring a pad-segmented cathode for 2-D readout, activated by a thin CsI layer, was effectively employed for pattern recognition in Ring Imaging Cherenkov counters. Even now (2025), these detectors are used in various experiments; for instance, the High Momentum Particle Identification detector in the last twenty years is collecting data in ALICE experiment at CERN's Large Hadron Collider. By the 1990s, MWPCs were central to nearly all experiments in particle physics.

In 1992, Georges Charpak was awarded the Nobel Prize in Physics "for inventing and developing particle detectors, especially the multiwire proportional chamber."

In 1995, Charpak created his final invention – a micropattern gas detector called MICROMEGAS. Presently, its altered edition is implemented in the ATLAS experiment at CERN.

The MWPC was the first electronic imaging device. Following that, additional varieties of imaging detectors were introduced: Microchannel Plate photomultiplier, position-sensitive solid-state detectors, position-sensitive photomultipliers, and gaseous micropattern detectors. They rely on the identical 2-D or 3-D signal readout principles that employ segmented electrodes and drift time, as initially established by G. Charpak's team and their successors.

At present, conceptually novel particle detectors are proposed for the High-Luminosity Large Hadron Collider (2025 onwards) and/or the Future Circular Collider. They consist of entirely solid-state solutions, 25 microns thick, featuring expansive silicon sheet surfaces measured in square meters that can be curved into cylindrical forms. Their functioning ought to be easier than that of gas detectors, featuring outstanding timing capabilities; they come with readout systems containing billions of channels built directly into the silicon detector. This advancement in technology appears to be setting the stage for substituting the dependable, traditional MWPC with conceptually innovative detectors in particle physics.

This book primarily seeks to highlight the creative application of the MWPC in particle physics, nonetheless, a review of its different uses in fields such as astrophysics, medicine, biology, industry, homeland security and environmental protection illustrates the importance of this "brilliant concept" beyond particle physics alone. A remarkable spin-off effect!

The latest advancements in modern electronic components and state-of the art signal processing techniques used in particle detectors are also discussed. Currently, Graphics Processing Units and Artificial Intelligence technologies are consistently enhancing the MWPC's performance in complex experiments at colliders.

This book is designed to appeal to a broad audience, encompassing a diverse range of readers, from students who are just beginning their exploration of the subject, to young researchers and engineers seeking to deepen their understanding. In line with this approach, the material presented spans a wide spectrum of complexity. Some chapters and sections are crafted with an accessible style and straightforward explanations, making the concepts easy to grasp for those new to the field. However, the book also includes more advanced material that delves into the technical intricacies of instrumentation and the fundamental principles of particle physics. These more demanding sections are intended to challenge readers who already possess a foundational knowledge in these areas, offering them the opportunity to expand their expertise and engage with the subject at a deeper level

In the end, we aspire for this book to motivate students and researchers to engage in detector development, exercising their creativity to advance experimental physics and science in general, like G. Charpak!

CHAPTER 1

A BIT OF DETECTOR HISTORY

Abstract

In this chapter, a brief history of the development of position-sensitive detectors for elementary particles and X-ray photons is provided. To fully recognize Charpak's impact on science and technology, it is crucial to examine the condition of particle detectors before and after 1968, the year he invented the MWPC.

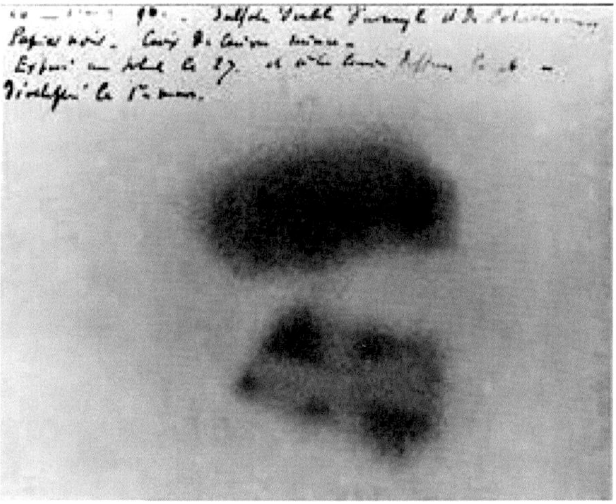

Figure 1-1. An image captured using a photographic plate wrapped in black paper and placed in contact with uranium salt ($K_2UO_2(SO_4)_2$). The shadow of a metal cross is visible, created by the mysterious radiation emitted by the uranium (from https://www.edpsciences.org/images/stories/archives/Becquerel.pdf).

Historically, the earliest "visualizations" of photons and elementary particles were achieved using film plates. For example, Fig. 1-1 shows an image of invisible radiation, which Becquerel accidentally discovered while studying phosphorescence.

Photographic plates were widely used from the late 19th century to the early 20th century by many researchers, including Röntgen, who utilized them to study, and later apply, the X-rays he discovered. For instance, Fig. 1-2 shows the famous X-ray image of his wife's hand, which marked the beginning of groundbreaking advancements in X-ray applications.

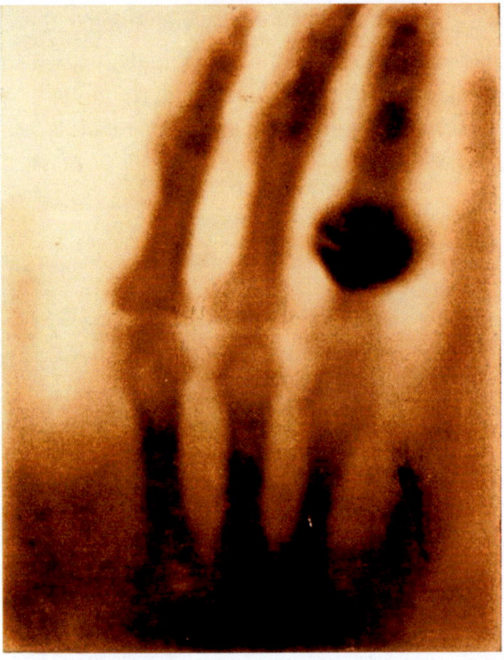

Figure 1-2. The X-ray image of Mrs. Wilhelm Röntgen's hand, captured using X-ray technology (from https://www.google.ru/search?q=images+obtained+by+roentgen&newwindow=1&tbm=isch&tbo=u&source=univ&sa=X&ved=0ahUKEwijlo7-0Z7bAhXrF5oKHb9iDcQQsAQIKQ&biw=1181&bih=847#imgrc=p60OGK5UiAiO6M),

At that time, however, the low sensitivity of available photographic films made it impossible to detect a single photon or elementary particle. This capability emerged a few years later, in 1908, with the invention of the Rutherford-Geiger counter.

1.1. Ionization chambers

The ionization chamber, historically the first gas detector used in measurements at the beginning of the 20th century, could adopt various geometrical forms, such as flat, cylindrical, or spherical, depending on experimental requirements. Despite these variations in shape, its operating principle remains unaffected by its geometry.

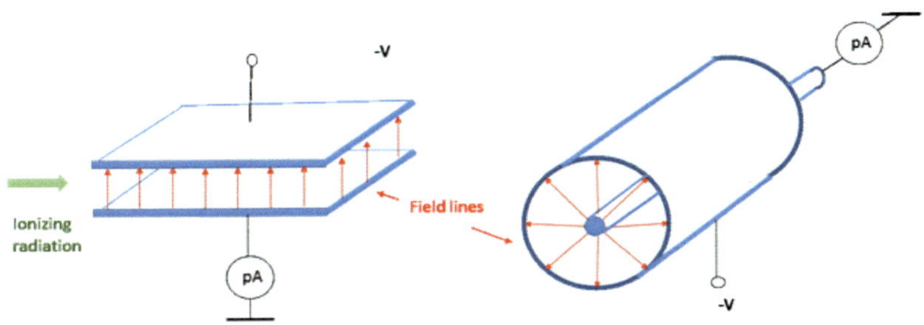

Figure 1-3. Schematic diagrams of planar and cylindrical ionization chambers

Fig. 1-3 provides a schematic representation of a planar and a cylindrical ionization chamber. These chambers consist of two metal electrodes, an anode and a cathode, with a voltage difference applied between them.

These detectors can operate in various gases, including air, typically at pressures around 1 atm. They remain widely used today, particularly in high-energy physics and medicine for dosimetry applications (see, for example, [https://en.wikipedia.org/wiki/Ionization_chamber#:~:text=Ion%20chambers%20are%20widely%20used,energies%20above%2050%2D100%20keV] and references therein).

When an intense flux of ionizing radiation (such as X-rays, gamma rays, or high-energy particles) enters the detector volume, it generates ion-electron pairs in the gas. The resulting current, measured as a function of the voltage applied between the electrodes, exhibits a characteristic behaviour:

- At low voltages, the current increases with voltage until it reaches a saturation region, often referred to as the 'plateau.'
- In this region, nearly all the ion-electron pairs produced by the ionizing radiation are collected on the electrodes before recombination can occur.

From the value of the saturation current, one can deduce the absolute intensity of the incident radiation.

1.2. Geiger-Muller counters

At the beginning of the 20th century, ionization chambers were widely used to study cosmic radiation. However, they lacked the ability to detect single events, a capability that became possible slightly later. In 1903, a ZnS scintillator was introduced into experimental techniques. This scintillator emitted a faint light pulse during ZnS de-excitation when ionizing particles passed through it. For the first time, this allowed the detection of individual particles.

This breakthrough was quickly adopted by Rutherford in his seminal experiments with alpha particles. At the time, the scintillation light produced by the ZnS screen could only be recorded visually, making the experiments labour-intensive and requiring significant preparation. Before measurements, researchers had to spend several hours in complete darkness to achieve the necessary visual sensitivity to detect the faint light pulses.

H. Geiger, a close collaborator of Rutherford, was an active participant in these experiments. Likely, the tedious nature of visually detecting particles inspired Geiger to seek a more efficient detection method. This led to the development of a groundbreaking idea: utilizing electron avalanche multiplication in gas to detect particles. This phenomenon had been discovered and studied a few years earlier by Townsend, as detailed in Chapter 1.

After conducting a series of experiments with Rutherford, Geiger successfully developed a reliable device in 1908 that could electronically detect single ionizing events (see Fig. 1-4).

The counter consisted of a coaxial cylinder filled with gas at a pressure of 0.1–1 atm. The outer cylinder acted as the cathode, while a thin wire at the center served as the anode. Due to this geometry, a strong electric field was generated near the wire. When a charged particle passed through the detector's gas volume, it produced primary ionization, creating electrons and positive ions.

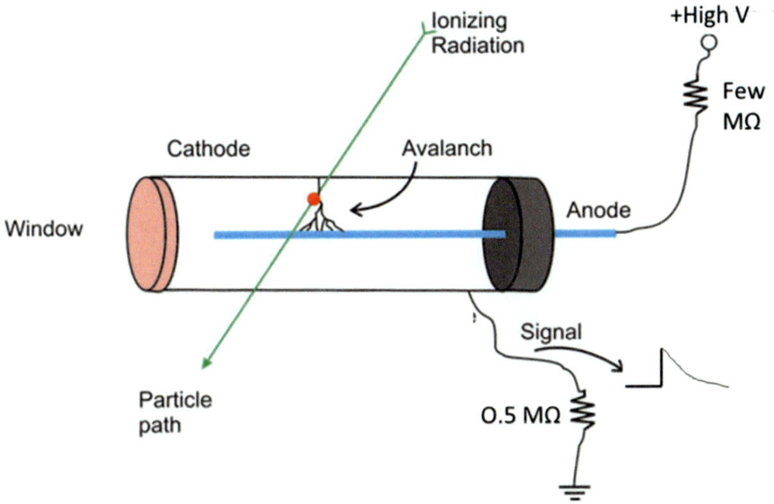

Figure 1-4. Schematics of Geiger counter (adapted from https://www.imagesco.com/geiger/geiger-counter-tube.html, with permission from John Iovine, Images Scientific Instruments Inc).

The primary electrons drifted toward the wire, triggering a corona discharge. This discharge generated a current, which produced a voltage pulse across a resistor load connected in series with the electrical circuit. The pulse amplitude was on the order of a few hundred volts, making it easy to detect. At the same time, the voltage drop caused by the pulse reduced the actual voltage across the detector electrodes, thereby interrupting the discharge.

After a short recovery time, typically on the order of milliseconds, the voltage was restored, and the detector was ready to record another event. This enabled the counter to detect up to 10^3 particles per second. This device was the first electronic detector of charged particles and can be regarded as a precursor to Charpak's later inventions.

1.3. Cloud chambers, bubble chambers and spark counters

At the beginning of the 20th century, photographic plates, scintillators, and Geiger counters were not imaging detectors of charged particles and lacked the ability to precisely determine the ionization tracks produced in the media. As a result, their contribution to particle physics was limited. The major advancements came with the next generation of detectors invented in the first half of the 20th century: cloud chambers, bubble chambers, and spark counters. These detectors enabled the visualization of particle tracks in three dimensions with high positional resolution, often much below 1 mm, providing physicists with powerful tools for their studies. For their groundbreaking contributions, the inventors of the first particle imaging detectors, Charles Wilson (cloud chamber, 1927) and Donald Glaser (bubble chamber, 1960), were awarded Nobel Prizes in Physics.

It is fascinating to explore how these brilliant individuals arrived at their inventions.

Charles Wilson originally focused on studying weather phenomena and sought to recreate cloud formation under laboratory conditions. To achieve this, he built a closed vessel filled with air saturated with water vapour (see Fig. 1-5). By rapidly expanding the volume of the vessel, he successfully created a supersaturated state of the vapour and observed that dust particles acted as condensation centers, playing a crucial role in the cloud formation process. To confirm this hypothesis, he meticulously cleaned the air used in his experiments to remove any dust particles.

To his great surprise, Wilson observed that even after removing all dust from the chamber, a small number of droplets still formed spontaneously. Moreover, these droplets often traced distinct paths, leading him to hypothesize that ions in the air could serve as condensation nuclei. To test this hypothesis, he exposed his device to the recently discovered X-rays and observed intense vapour condensation. Shortly thereafter, Wilson also reported observing alpha, beta, and gamma rays. This marked the development of the first imaging detector of ionizing radiation!

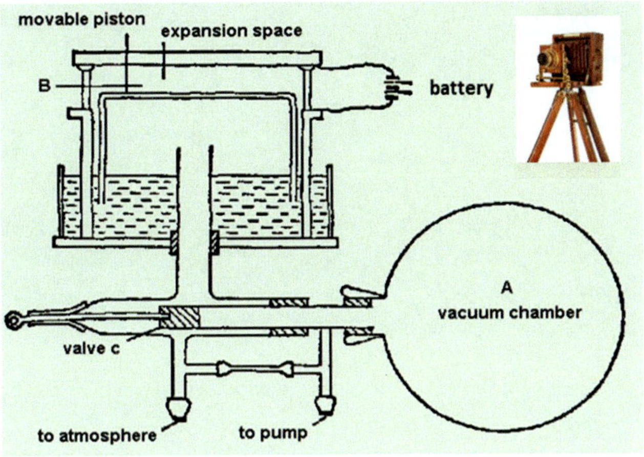

Figure 1-5. One of the original drawings of Wilson's chamber. The water-saturated air is contained in the volume marked B, while the vacuum vessel, labeled A, is kept evacuated with the valve C closed. To create supersaturated vapours, valve C is opened, allowing the air under the piston to be pumped into vessel A. As the piston falls, it causes the air in B to expand and cool, enabling the formation of droplets at centers of condensation (adapted from: https://www.uefap.com/reading/exercise/ess3/campbell.html).

A photographic technique was employed to record these tracks, producing their 2D images. Examples of such images are shown in Figs 1-6 and 1-7. The spatial resolution of this detector was approximately 0.5 mm. ([Brander, H "Bubble chambers"_ http://laplace.ucv.cl/Cursos/Fisica_Contemporanea_2/Old/annurev.ns.10.120160.000545.pdf) The expansion cycle frequency, in the best cases, ranged from 0.2 to 0.5 Hz, although, in principle, it could be even higher [Gaerther, 1949, 588-595], [Curtiss, 1933, 579-58].

Wilson imaging detectors made a remarkable contribution to high-energy physics experiments. For example, they played a crucial role in the discovery of several elementary particles, including the positron (1931), the muon (1936), and the kaon (1947).

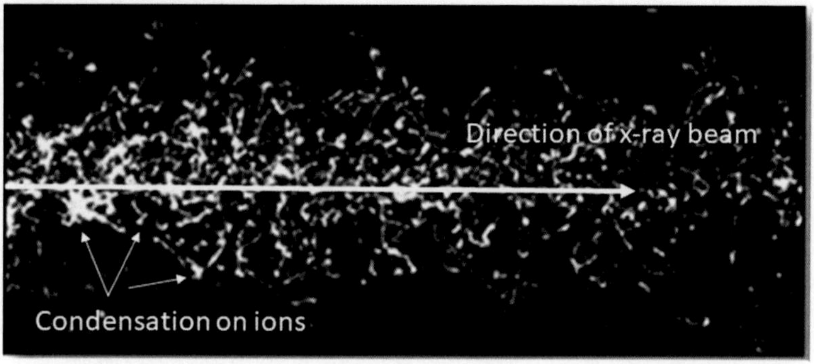

Figure 1-6. Images of ionization clusters produced in the Wilson chamber by X-ray radiation (adapted from: http://www.cambridgephysics.org/cloudchamber/cloudchamber13_1.html)

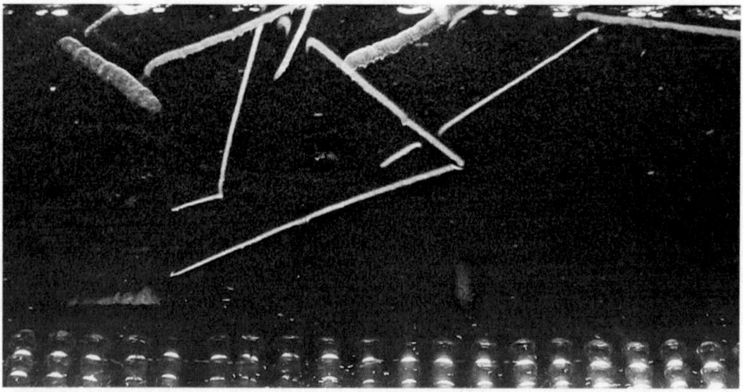

Figure 1-7. Images of ^{220}Rn decay events captured in a cloud chamber (source: https://commons.wikimedia.org/wiki/File:Radon220_decay_in_a_cloud_chamber.jpg, licensed under CC BY-SA 3.0.).

However, like any device, cloud chambers had their limitations. They were too small for use in large accelerators, and the air density was insufficient to effectively interact with highly energetic particles. Additionally, the piston cycle of the chambers was too slow to keep up with the rapidly increasing speed of accelerator cycles.

These drawbacks were partially addressed in 1952 with the development of another imaging detector by Donald Glaser. He successfully achieved bubble formation on ions in superheated fluids, paving the way for a new approach to particle detection.

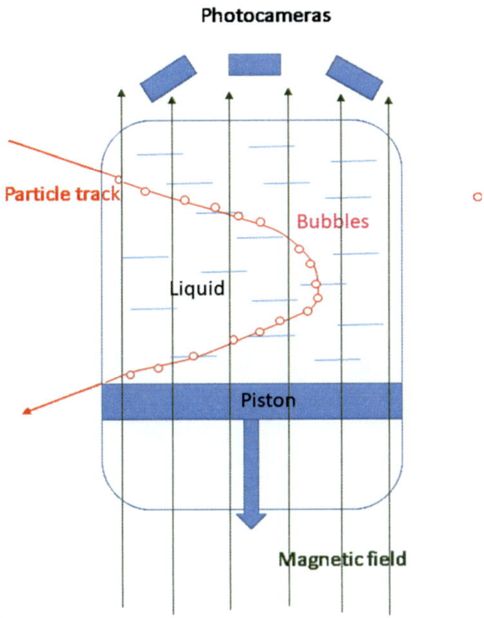

Figure 1-8. Schematic illustration of the Glaser bubble chamber.

The bubble chamber, in principle, was somewhat similar to the Wilson cloud chamber. It consisted of a large cylinder filled with a liquid heated just below its boiling point (Fig. 1-8). At the bottom of the chamber, a piston was installed. When a charged particle passed through the chamber, the piston rapidly reduced the internal pressure, causing the liquid to transition into a superheated phase. This process led to the formation of bubbles along the particle's path, making its trajectory visible.

To record stereoscopic images of these tracks, several cameras were mounted around the chamber. The typical position resolution of such a detector was approximately 50 μm, and the expansion cycle frequency was about 10 Hz (Brander, H., Bubble chambers_ http://laplace.ucv.cl/Cursos/Fisica_Contemporanea_2/Old/annurev.ns.10.120160.000545.pdf)

Typically, the bubble chamber is placed inside a solenoid, where a strong magnetic field causes charged particles to move in helical trajectories. By analyzing the radius of these trajectories, one can determine the charge-to-mass ratio and the velocities of the particles (see, for example, Fig. 1-9).

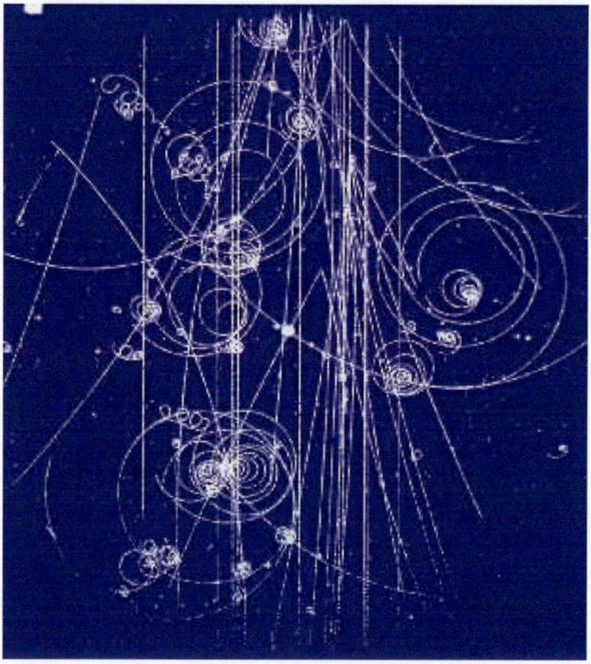

Figure 1-9. Example of particle track images obtained using a bubble chamber. Image credit: CERN.

With the help of bubble chambers, significant discoveries were made. For instance, in 1973, the Gargamelle experiment at CERN revealed *weak neutral currents*, confirming the possibility of unifying the electromagnetic and weak interactions within the electroweak theory. Remarkably, bubble chambers are still employed in some experiments today, such as in PICO (Project in Innovative COld technology) [https://indico.cern.ch/event/868940/contributions/3814879/attachments/2082697/3498379/PICO_ICHEP_Eric_VJ.pdf]

Despite the triumph of the cloud and bubble chambers, they soon became too "slow" to satisfy the growing requirements of high-energy physics experiments. This is why the attention of researchers turned towards another type of detector: spark chambers. First, such a detector consisted of two parallel-plate metallic electrodes with a distance between them around a few mm [Keuffle, 1949, 202-208], [Bella, 1953, 1338-1340], and [Bella, 1953, 1461-1479]. If a sufficiently high voltage was applied between the electrodes, then a particle, traversing the chamber, may cause there a well-localized spark. The spark was photographed from two perpendicular directions, which allowed one to determine its 2D coordinate with an accuracy of about 1 mm. In some designs, an external electronic circuit was used to turn out the spark and make it read after a few ms to record the next event.

The early versions of the spark chamber had several technical drawbacks and, for this reason, were primarily used in academic studies. Furthermore, since each spark was mainly initiated by primary electrons created near the cathode, its trajectory did not accurately represent the particle track, posing a fundamental limitation to track reconstruction.

The advanced version of the spark chamber (see [Trumper, 1960. 121-128]), which challenges the cloud and bubble chambers and became widely used in experiments in the early 1960s, is shown in Fig. 1-10.

This detector consisted of a stack of parallel plate counters, with a relatively low voltage (approximately 1 kV) applied across each gap. It was surrounded by triggering devices, such as Geiger counters or scintillators. When a charged particle passed simultaneously through two planes of these triggering devices, a short (~1 μs) high-voltage pulse (≥ 10 kV) was applied to the parallel-plate chambers, causing sparks to form within them. These sparks appeared simultaneously across the stack of parallel-plate chambers,

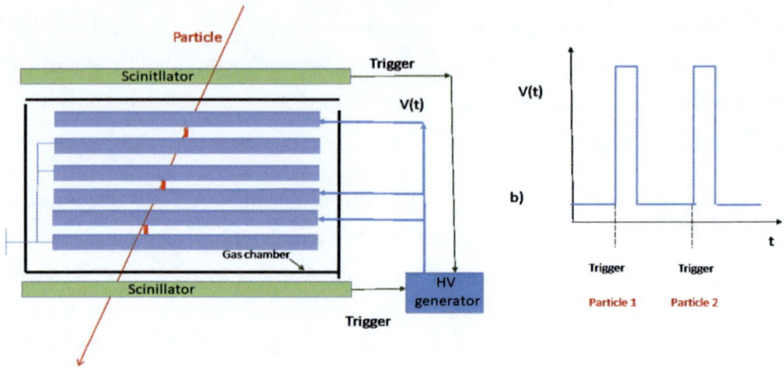

Figure 1-10. Schematic diagram of a pulsed spark chamber.

roughly aligned along the particle's track (see Fig. 1-11), enabling imaging capabilities.

By employing mesh electrodes and a system of photo cameras, it was possible to record 3D images of the particle tracks. The primary advantage of this new detector was its significantly higher repetition rate for the active phase (pulsed voltages), which was nearly 100 times greater than that of the bubble chamber.

The large-volume spark chambers played a pivotal role in the discovery of the muon-type neutrino in 1962.

A further advancement of the spark chamber led to the development of the so-called streamer chamber. This unique wide-gap chamber utilized a much shorter high-voltage (HV) pulse, lasting only a few nanoseconds, compared to the spark counter (Fig. 1-12). Primary electrons, generated by the passage of charged particles, initiated the formation of streamers. However, due to the brevity of the HV pulse, the streamers produced by each electron cluster along the particle's track propagated only a short distance before decaying when the voltage was abruptly removed. Each of these streamers emitted a burst of light intense enough to visualize the particle track.

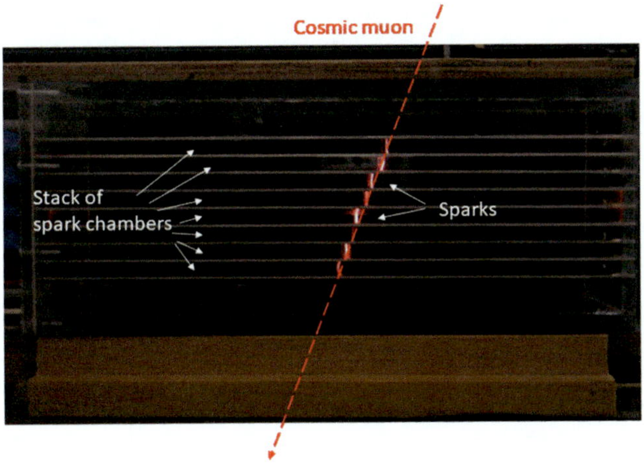

Figure 1-11. Photograph showing sparks appearing simultaneously in a stack of pulsed spark chambers along the track of a charged particle (adapted from *http://www.hep.phy.cam.ac.uk/~lester/teaching/SparkChamber/Gallery.html).*

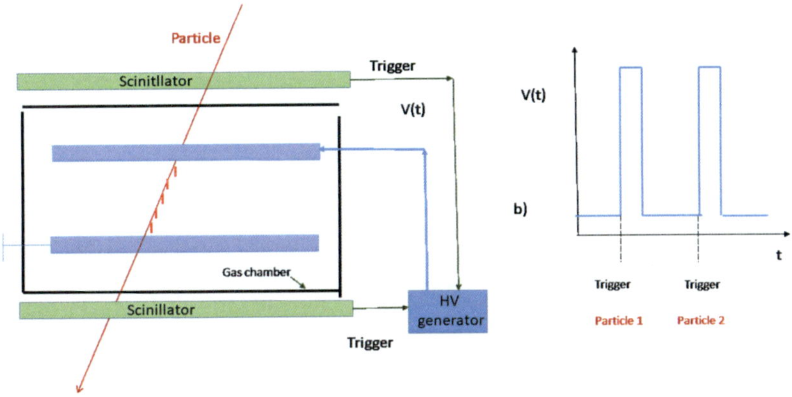

Figure 1-12. Schematic drawing illustrating the principle of operation of a steamer chamber.

Let us emphasize once again that in all the types of imaging detectors mentioned above (cloud chambers, bubble chambers, spark chambers, and streamer chambers) a photographic technique was employed to record particle tracks within their fiducial volumes. Until the 1970s, spark and streamer chambers represented the cutting-edge of particle detection technology. During this period, Wilson and Glaser's chambers gradually fell out of use in experimental setups

1.4. Multiwire proportional chambers and their descendants

The next major breakthrough occurred in 1968 when G. Charpak, who had previously worked with spark chambers, invented the multiwire proportional chamber (MWPC) [Charpak, 1968, 262-268] - see Fig. 1-13.

The first version of the MWPC was remarkably simple in design. It consisted of two parallel-plate cathodes, with thin, stretched anode wires positioned at the center of the gap (Fig. 1-14).

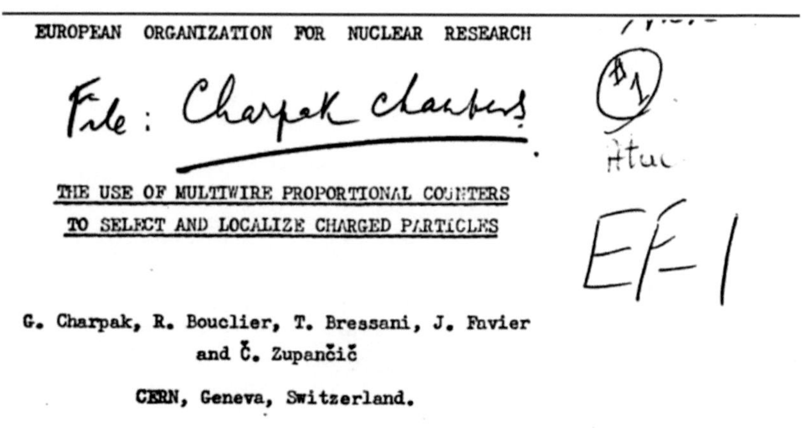
EUROPEAN ORGANIZATION FOR NUCLEAR RESEARCH

File: Charpak chambers

THE USE OF MULTIWIRE PROPORTIONAL COUNTERS
TO SELECT AND LOCALIZE CHARGED PARTICLES

G. Charpak, R. Bouclier, T. Bressani, J. Favier
and Č. Zupančič

CERN, Geneva, Switzerland.

EF-1

Figure 1-13. Photocopy of the first page of the CERN preprint describing the multiwire proportional chamber, featuring Charpak's handwritten notes. CERN image.

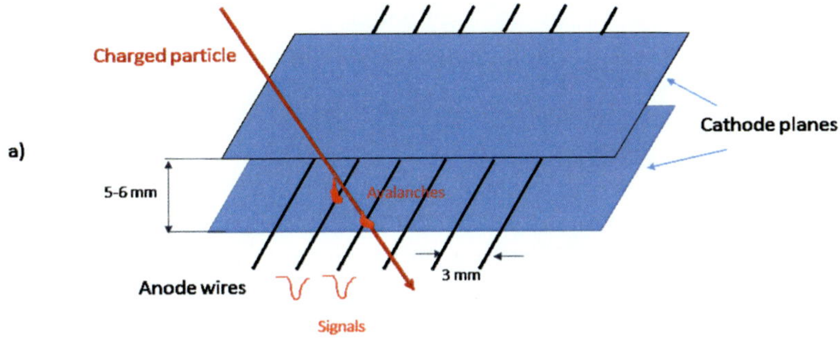

Fig. 1-14. Conceptual diagram of the multiwire proportional chamber (MWPC).

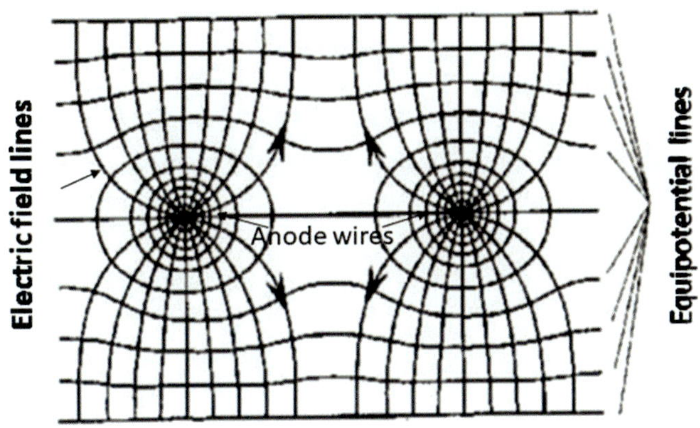

Figure 1-15. Electric field line map of a multiwire proportional chamber (MWPC), illustrating the field distribution around the anode and cathode wires (adapted from [Charpak, 1970, 13-34]).

The electric field line map shown in Fig. 1-15 illustrates that the MWPC functions as an array of single-wire counters, each with an independent fiducial volume. Primary electrons generated within these volumes drift toward the nearest wire, where they produce avalanches. Each wire is connected to its own charge-sensitive amplifier, enabling the determination of which fiducial volume the charged particle traversed and its ionization energy loss. In essence, the signal amplitude is proportional to the number of primary ions created within the specific sub-volume.

The unique features of this new detector can be summarized as follows [Charpak, 1968, 262-268]

- Each wire can amplify the initial energy loss of a particle in a thin gas layer (approximately 1 cm) to such an extent that minimum ionizing particles are detected with an efficiency close to 100%.
- In certain gas mixtures, the charge amplification factor can reach values as high as 10^5-10^6, facilitating the detection and processing of electronic signals.
- A counting rate of up to 10^5 events per second per wire can be achieved, enabling the detection of high event rates, such as those produced in accelerator experiments.
- The time resolution of the MWPC is below 0.4 μs.
- Localization of the primary charge position between wires is possible by utilizing the arrival time of the pulse.
- The MWPC allows for particle identification through the selection of particles with different ionization powers.
- MWPCs can operate effectively in strong magnetic fields.

A stack of such detectors, with wires oriented perpendicular to each neighbouring plane and combined with computer-based signal processing, enabled the fast generation of 3D electronic images of particle tracks with a position resolution of approximately 1 mm. For the first time, particle tracks could be recorded rapidly and without relying on photographic techniques. This marked a revolutionary advancement in particle detection technology.

G. Charpak often joked that he had invented an “Electronic Bubble Chamber” to highlight the transformative nature of this innovation.

To reduce the detector mass, MWPCs with cathode planes made of wires were soon developed (see, for example, [Dimcovski, 1971, 151-155]).

Shortly thereafter, further studies revealed that the induced signals from these wires (see Fig. 1-16) could be used for accurately determining the position of primary ionization (see, for example, [Charpak, 1979, 455-464] and references therein).

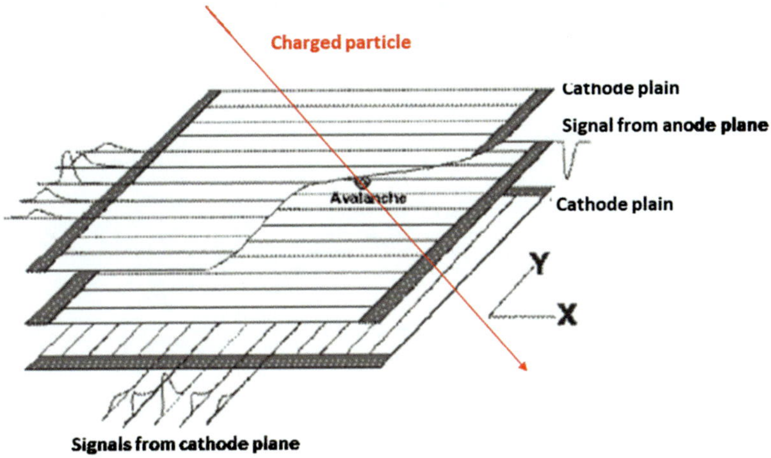

Figure 1-16. Schematic illustration of signals induced by an avalanche on surrounding wire (adapted from: file:///C:/Users/Win7/Downloads/phd-tianbowang.pdf).

The significance of this approach was fully appreciated only later, as it not only enabled a considerable improvement in the spatial resolution of MWPCs but also found applications in other types of detectors. These include solid-state detectors, vacuum photodetectors, certain liquid time projection chambers, and micropattern gaseous detectors. By applying a specialized algorithm for signal processing (centre of gravity, see Chapter 3), a position resolution of 35 μm along the anode wires and 150 μm in the perpendicular direction was achieved using soft X-rays. Both values represent the distribution sigma.

As an example, Fig. 1-17 shows an X-ray radiograph of the word "CERN," obtained with an MWPC featuring a 2 mm wire pitch [Charpak, 1978, 471-482]. A copper mask containing the word was positioned in front of the MWPC, parallel to it, with the longer dimension of the word CERN aligned parallel to the anode wires.

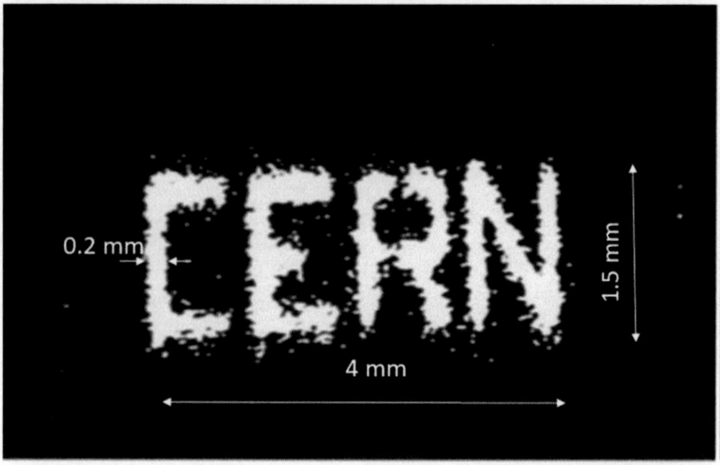

Figure 1-17. Two-dimensional X-ray image of a copper mask with a cut-out letter (adapted from [Charpak, 1978, 471-482]).

This and other electronic images obtained with MWPCs greatly excited the high-energy physics community and triggered a series of rapid advancements in the field.

Almost immediately after, at the end of the 1970s, two significant inventions were made: photosensitive MWPCs and MWPCs with pad-segmented cathodes. These advancements enabled, for example, the detection of Cherenkov light cones, facilitating particle identification within specific momentum ranges. This innovative detector became known as the Ring Imaging Cherenkov detector (RICH, see Chapter 5).

Around the same time, the drift chamber (see Chapter 5) was developed, followed shortly by the Time Projection Chamber (TPC). Both detectors were based on MWPC technology and allowed for the reconstruction of three-dimensional particle tracks with impressive precision (see Chapter 5). By measuring the curvature of these tracks and the ionization density along them, highly effective particle identification became possible.

These developments mark the beginning of a new era in detection techniques. MWPCs have enabled physicists to make significant discoveries in high-energy physics, such as the W and Z bosons (see Chapter 10).

Further progress has been made possible by the rapid development of fast electronics for signal processing and advanced online data-handling techniques utilizing Graphics Processing Units. More recently, Artificial Intelligence programs have played a critical role in analyzing the vast amounts of data generated by experiments, such as those conducted at the CERN LHC. A notable example is the ALICE experiment, where a TPC based on MWPC technology was in use until 2018.

Today, nearly every experiment in particle physics employs wire-based proportional detectors. Examples include TPCs, muon tracking chambers, transition radiation detectors, RICH detectors, and even readout components in some calorimeters.

1.5. Other imaging detectors

In the late 1960s, solid-state detectors began to develop rapidly. These detectors exhibit an absorption coefficient, or "stopping power," for ionizing radiation that is typically orders of magnitude greater than that of gas or liquid detectors. Furthermore, for the same amount of absorbed ionizing radiation energy, the number of primary electrons generated in a solid-state detector is nearly ten times greater than in a gas or liquid detector.

This characteristic allows solid-state detectors to operate effectively in the so-called "pulse ionization chamber mode," where, instead of recording a steady current, the detector captures pulses produced by individual particles or X-ray photons.

To deviate slightly from the chronological development of solid-state detectors, this chapter will focus on describing the simplest design of such a pulse chamber. A notable example of a detector of this type is the widely used CdZnTe detector. For further details, see, for example, [Del Sordo, 2009, 3491-3526] and references therein (see also Fig. 1-18).

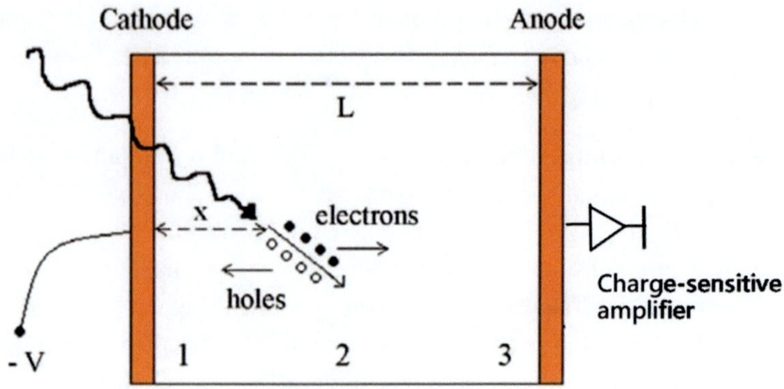

Figure 1-18. Schematic drawing of a CdZnTe detector. This detector functions as a solid-state ionization chamber operating in pulse mode: drifting primary ions generate a current, and the time-integrated current, representing the charge, is recorded by a charge-sensitive amplifier (adapted from [Del Sordo, 2009, 3491-3526]).

Although CdZnTe is a semiconductor, it has a relatively high resistance. As a result, when a voltage is applied through specially designed electrodes, the leakage current remains negligible. However, when ionizing radiation generates electron-hole pairs within the CdZnTe, these charges can drift freely under the influence of the applied electric field, producing a current pulse. In this respect, the detector functions similarly to a gas or liquid ionization chamber.

To enhance sensitivity, it is common to measure not the current pulse itself but its time integral, the total charge collected on the electrodes.

In the case of segmented electrodes (Fig. 1-19), an approach inspired by Charpak's work on MWPCs, moving primary charges induce signals on the electrodes, similar to the process in a wire chamber. By analysing the signal distribution, which depends on the geometry of the segmented electrodes, it becomes possible to determine the one-dimensional or two-dimensional coordinates of the primary electron cloud.

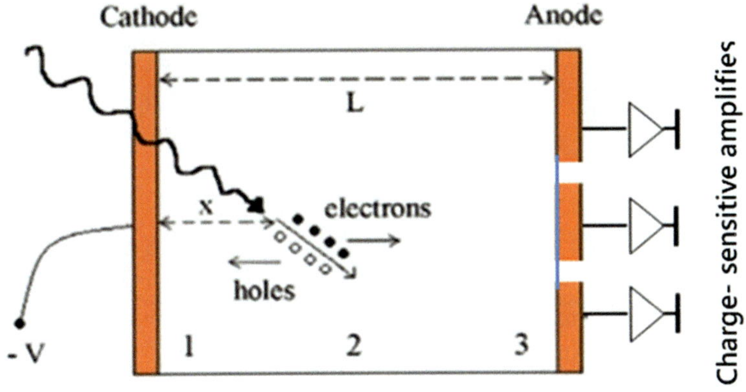

Figure 1-19. CdZnTe detector with a segmented anode. The distribution of induced charges allows the determination of the one-dimensional or two-dimensional coordinates of the primary electron cloud (adapted from [Del Sordo, 2009, 3491-3526]).

CdZnTe detectors are primarily used for X-ray and gamma imaging. In high-energy physics, silicon (Si) detectors are widely employed for particle tracking. These detectors operate in pulse ionization chamber mode. However, due to the relatively high conductivity of silicon, a p-n junction is used to create a depletion region, which acts as a high-resistivity zone.

To enhance detector performance, a high voltage is applied in reverse bias (negative to the p-side, positive to the n-side). This expands the depletion region, reducing leakage current and increasing sensitivity.

When a high-energy particle passes through the detector, it ionizes silicon atoms, generating electron-hole pairs within the depletion region. The electric field in this zone swiftly separates these charges, producing a measurable electrical pulse.

Solid-state detectors have challenged MWPCs in several applications. While MWPCs remain widely used in areas requiring large-area coverage and lower costs, solid-state detectors are now the preferred choice in many fields due to their superior position and energy resolution, compactness, and advanced capabilities. They are particularly dominant in fields such as medical imaging, X-ray spectroscopy, and high-energy physics experiments that demand precision detection.

In the early 1980s, the interest of physicists turned toward resistive plate chambers, which were first tested by the Pestov group in Novosibirsk and further developed by Italian physicists led by R. Santonico and R. Cardarelli. These detectors use parallel plates made of resistive materials, such as glass or Bakelite, with a gas gap to detect charged particles through ionization. Their main feature is that they are spark-protected, allowing them to operate safely at high gas gains and achieve, depending on the specific design, time resolutions ranging from 50 picoseconds to 1 nanosecond.

Introduced in the late 1980s, micropattern gaseous detectors are more compact than MWPCs, and some of their characteristics are approaching those of solid-state detectors. In the original design invented by A, Oed [Oed, 1988, 351-59], the anode and cathode wires were replaced with metallic microstrips on flexible or standard Printed Circuit Board (PCB) substrates. Later, G. Charpak [Charpak, 1995, 1-30] and collaborators developed a microgap parallel-plate avalanche chamber, known as MICROMEGAS, while F. Sauli introduced a microhole gas amplification structure –GEM [Sauli, 1997, 531-534].

Today, these detectors are considered viable alternatives to wire chambers in certain applications. Their reduced electron cluster diffusion (due to the shorter drift distances, strong, well-controlled electric fields, and fine segmentation of the amplification and readout structure) enables better spatial resolution in tracking devices. However, since the underlying physics governing their operation remains the same as that of wire chambers, a strong understanding of wire chamber technology is essential for the development of micropattern detectors. Therefore, a comprehensive book covering all aspects of wire chambers, including associated signal readout methods, would be valuable not only for researchers working on wire chambers but also for those developing micropattern detectors.

Micropattern detectors represent the second generation of an "electronic bubble chamber." The development of multiwire chambers has had a profound and lasting impact on detector technology, with its influence still evident today. It is important to reflect on this evolution in the book, providing readers with a clear understanding of the current state of particle detectors and potential future developments.

However, the primary goal is to illustrate how a simple, yet brilliant idea can transform the field of particle and photon detection. This serves as an instructive example of how scientific and technological progress often occurs through sudden breakthroughs, driven by innovative ideas.

Concluding Remarks

As demonstrated in this chapter, detectors for elementary particles and X-ray photons have undergone remarkable advancements, driven by a series of breakthrough developments over the past century and into the present. Among these innovations, key contributions such as MWPCs, their readout techniques, and MICROMEGAS were pioneered by Georges Charpak and his team, marking a significant milestone in the evolution of particle detection.

In the next chapter, we will explore some early classical experiments that laid the foundation for the invention of MWPCs.

CHAPTER 2

FROM PRECURSORS TO INNOVATION: CHARPAK'S LEGACY AND THE FUNDAMENTALS OF AVALANCHE GASEOUS DETECTORS

Abstract

In its initial configuration, the MWPC can be regarded as an array of single-wire gas counters. These gas counters, originally invented by Rutherford and Geiger, operate based on the principle of gas avalanche multiplication, an effect first discovered and studied by Townsend. This phenomenon is briefly described below.

In general, this chapter provides readers with the fundamentals of avalanche detectors, information that will be useful both for their education and for further reading of this book.

2.1. Townsend avalanche

In his classical experiments, Townsend used a glass vessel containing two parallel-plate electrodes (see Fig. 2-1). The electrodes were connected to a battery (plus a voltage trimmer) in series with a milliammeter. The distance between the electrodes could be adjusted from 0.5 to 5 cm. The vessel was irradiated with an X-ray tube, and the current between the electrodes was measured at various air pressures, ranging from 0.1 to a few millimetres of mercury (mm Hg).

Some characteristics curves voltage/cm-current are shown in Fig. 2-2. The V/cm is the strength of the Electric field E. As observed, at low E field, the current exhibits a "plateau" region (marked with a red line in the figure). In this region, nearly all ions generated during the ionization process are collected, some of them undergo recombination therefore only a fraction reach the electrodes.

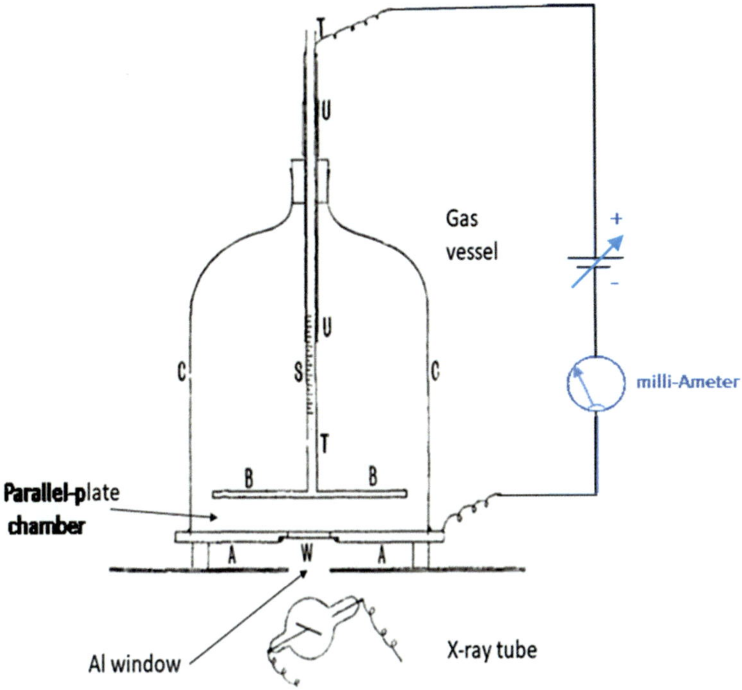

Figure 2-1. Schematic drawing of the setup used by Townsend in his experiments (adapted from [Townsend, 1901, 198-227]).

The operation of the detector with the value of E in the range from zero to the end of the plateau region, is commonly referred to as the "ionization chamber mode." Beyond this region, as the E increases further, the current begins to rise exponentially and can increase by at least an order of magnitude. The explanation proposed by Townsend (see [Townsend, 1901, 198-227]) assumed that X-rays ionize the gas, producing positive and negative (electron) primary ions. The latter are accelerated in an electric field, and if the field strength is sufficiently high, they can ionize neutral atoms and molecules in the residual air, generating secondary ions. Townsend mathematically described this process using the symbol *X* for the electric field strength; however, throughout this book, we will use the more commonly adopted symbol *E*.

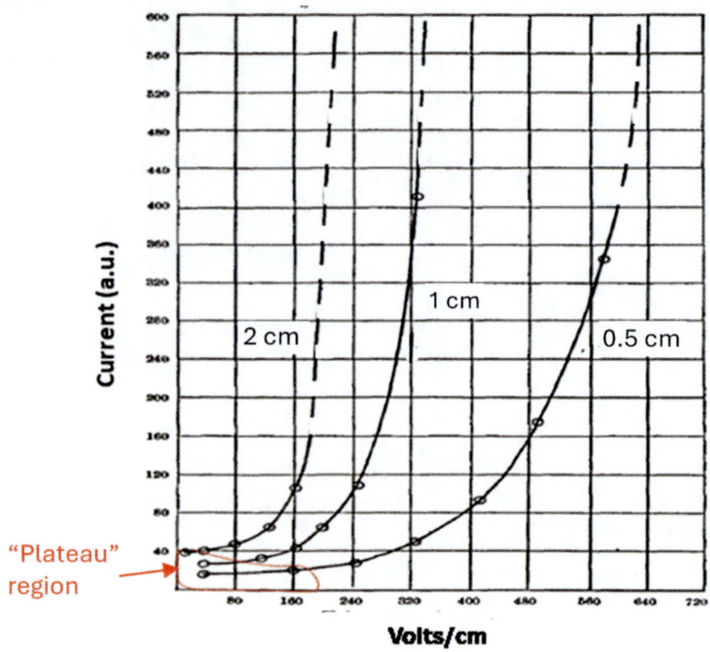

Figure 2-2. Current versus E (V/m) dependencies measured with the device shown in Fig. 2-1 at a pressure of 2.1 mm Hg. Numbers near the curves represent the values of the gap between the electrodes in cm (adapted from [Townsend, 1901, 198-227]).

He found that the coefficient α (later called the "First Townsend coefficient"), representing the number of ion pairs produced per unit length of drift, depends on X/p, where p is the gas pressure in the vessel.

> The new negative ions travel with the same velocity as the original N_0 ions, so that all the negative ions will be found together during the motion. The number of negative ions produced by N ions travelling through a distance dx will be $\alpha N dx$; where α is a constant depending on X, p, and t.
>
> Then
>
> $$dN = \alpha N dx.$$
>
> Hence
>
> $$N = N_0 \epsilon^{\alpha x}. \quad . \quad . \quad . \quad . \quad . \quad . \quad . \quad (1)$$

> The relation connecting the three variables α, p, and X was found to be of the form $\alpha = pf\left(\frac{X}{p}\right)$, which shows that α is proportional to p when $\frac{X}{p}$ is constant.

Later, the following expression was established for α:

$$\alpha = p\,A^{(-Bp/E)} \qquad (2\text{-}1)$$

where A and B are parameters that depend on the gas type and the electric field range. This expression provides a satisfactorily good approximation for the reduced First Townsend coefficient in many gases (see, for example, [Aoyama, 1985, 125-131]).

The first Townsend coefficient was measured for many gases used in practical devices (see Fig. 2-3). It can also be accurately calculated, for example, using the MATZDOLD program.

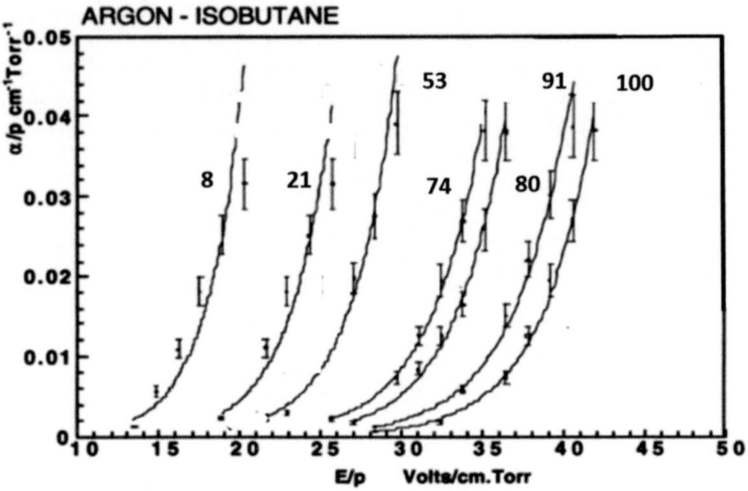

Figure 2-3. Example measurements of the first Townsend coefficient for various Ar-isobutane mixtures. Numbers in the plot indicate the isobutane percentage. Adapted from [Sharma, 1992, 280–283].

2.2. Geiger and proportional counter

A few years later, Rutherford and Geiger had two significant insights, another example of how bright ideas drive scientific progress, on how to take advantage of the gas multiplication phenomenon discovered by Townsend [Rutherford, 1908, 141-161:

1. In Townsend's experiments, avalanche multiplication was demonstrated in current mode. Rutherford and Geiger's first major insight was that the avalanche process could be exploited for detecting individual ionization events. A single particle would trigger avalanche multiplication, producing a distinct response.
2. Their second key idea was determining the best way to implement the first insight. Instead of using a parallel plate configuration, they proposed a cylindrical geometry (see Fig. 2-4). In this setup, the electric field varies with distance r from the center of the cylinder as 1/r, creating the strongest field near the anode wire. This fundamentally

changes the operating principle compared to the parallel-plate chamber: primary electrons generated in the detector volume first drift toward the anode wire without multiplication, then trigger the avalanche process in a very narrow region, only a few wire radii in diameter, where the electric field is strongest. Additionally, the increased drift volume in the cylindrical geometry enhances detection efficiency, as any primary ionization will lead to avalanche development and, consequently, particle detection.

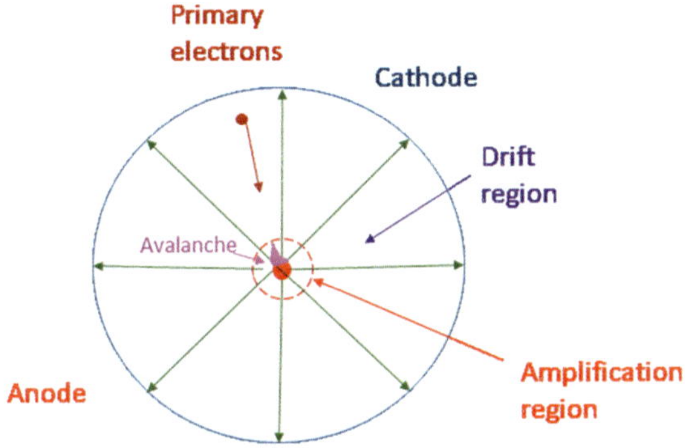

Figure 2-4. Schematic representation of the two main regions in a Geiger counter: the drift region (low electric field) and the multiplication region (high electric field).

One can calculate the number of secondary electrons produced in the avalanche region ($r < r_m$) by a single electron using the following formula:

$$A(r) = \exp \int_{ra}^{rc} \alpha(r)dr = \exp \int_{ra}^{rm} \alpha(r)dr \qquad (2\text{-}2)$$

where r_a, r_c, and r_m are the radius of the anode wire, the cathode, and the multiplication region respectively. The value of A is often referred to as the gas gain.

In general, an electron drifting between two points r_1 and r_2 in the field of this geometry will produce secondary electrons if its kinetic energy is sufficient to ionize an atom.

If n_0 primary electrons point-like cloud trigger avalanches, the total number of electrons produced will be:

$$N = n_0 A \quad (2\text{-}3)$$

In the Fig. 2-5, the current produced by a primary charge point-like cloud, created in the drift region of a cylinder counter by ionizing radiation, is shown. In this figure, t_{dr} represents the drift time, necessary for primary electrons to reach the amplification (or avalanche) region. During this time, the induced current is extremely small.

When the primary electrons approach the anode at a distance less than r_m, multiplication begins, and the current growing exponentially until the electrons are fully collected on the anode (red curve). For $t > t_{dr}+t_-$ the current is generated by the slowly moving positive ions in the electric field which decreases as a function of the drifting distance from the anode following 1/r dependence.

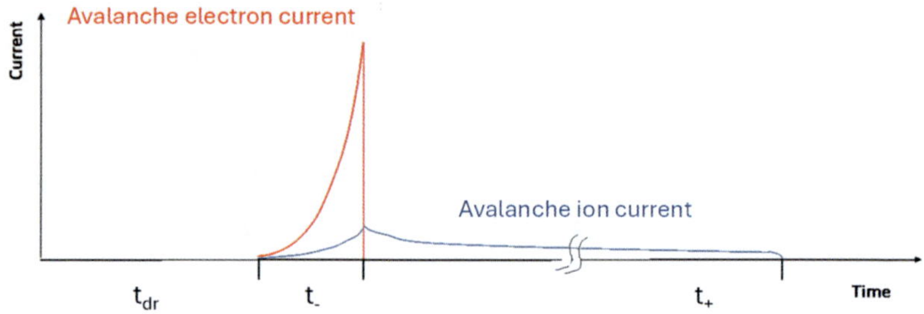

Figure 2-5. Schematic representation of the current induced on the electrodes of a cylindrical counter (not to scale). For illustration purposes, the figure assumes that the drift time of positive ions t_+ is approximately 10 times longer than the electron drift time t_-. However, t_+ is typically about 10^3 times longer than t_-.

In the proportional regime of operation, with a gas gain on the order of 10^3–10^5, the current produced by a single avalanche is typically small. For this reason, Charge-Sensitive Amplifiers (CSAs) are commonly used. These amplifiers generate an output voltage pulse with a height proportional to the integrated current (or equivalently, to the total charge in the avalanche) provided that the amplifier's integration time is sufficient for the ions to reach the cathode.

As an example, Fig. 2-6 shows schematically a typical charge signal from a single-wire counter recorded using a charge-sensitive amplifier. This signal represents the integral of Fig. 2-5, with the largest contribution coming from ions. This occurs because the charge induced by drifting ions is proportional to the voltage drop they experience while moving toward the cathode (from HV to zero volts).

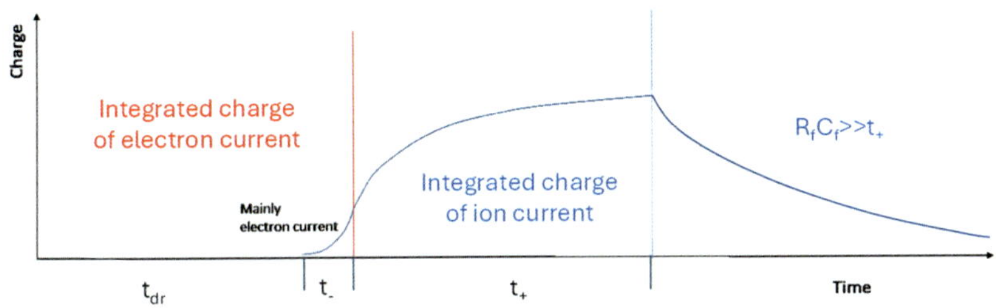

Figure 2-6. The typical charge signal shape from a single-wire counter.

In practice, electronic circuits known as "shapers" can be used to filter out the slow component of the signal while preserving the fast one. Fig. 2-7 shows the schematic of a single-wire counter equipped with a charge-sensitive amplifier (CSA) and a shaper based on a simple R_fC_f circuit and this time showing the corresponding V_{out}.

In Fig. 2-8, the time evolution of the CSA output voltage is schematically shown in blue. Typically, the output signal is shortened using a shaper (green curve, labelled "Output signal from the shaper"), which worsens the signal-to-noise ratio by removing a significant portion of the total signal. Additionally, fluctuations in the small integrated current degrade the detector's energy resolution.

Moreover, while the electronics can accept new pulses, the ion avalanche from the previous event is still migrating toward the cathode within the detector. At very high event rates, space charge effects can disrupt the detector's linear behaviour, further reducing its energy resolution.

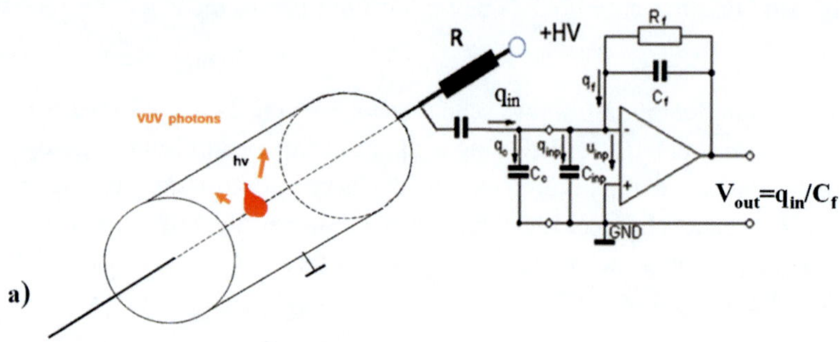

Figure 2-7. Schematic of a single-wire counter connected to a charge-sensitive amplifier with a shaper. The shaper is simplified as an R_fC_f time constant (ranging from hundreds of nanoseconds to microseconds), which limits V_{out} at a defined duration.

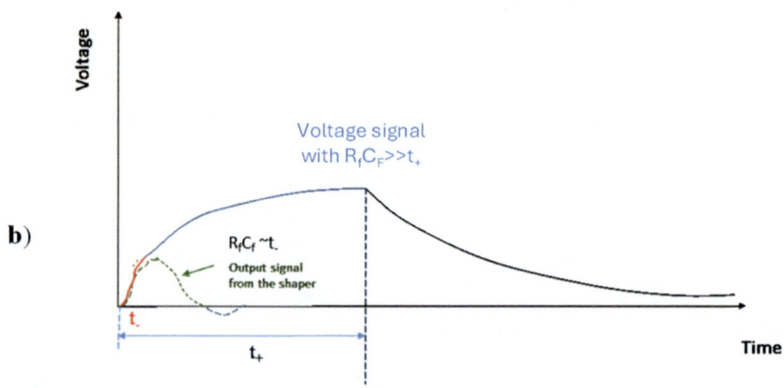

Fig. 2-8. The blue curve represents the voltage output signal from the charge-sensitive amplifier with R_fC_f>> t_+, whereas the green dashed line depicts the shaped signal after suppressing the slow component caused by the drifting of positive ions ($R_fC_f \sim t_-$).

To suppress the undershoot observed in the green curve, additional stages may be added to the electronic channel to rapidly restore the baseline ("zero line"), preventing bias in subsequent measurements of the pulse-height distribution.

Fig. 2-9 shows an example of the real signal measured at the output of the CSA using a digital oscilloscope. The signal consists of a short, sharp peak followed by a relatively long tail lasting a few microseconds. The exact shape of the pulses and their timing characteristics depend on the shaper bandwidth settings. In this example, the peak width is approximately 1 μs.

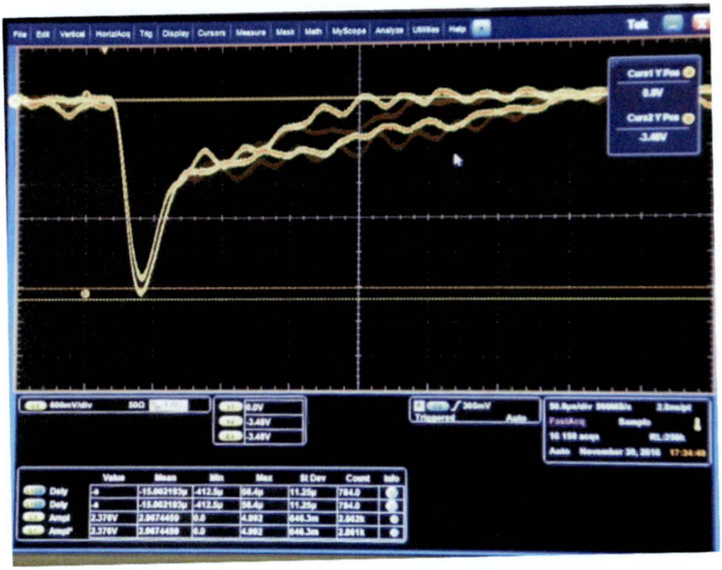

Figure 2-9. Typical pulse shape from a single-wire counter filtered by the Ortec 873 shaper. The fast signal is clearly visible, along with a tail resulting from the slow component.

For comparison, when considering the signal fraction over such a short pulse, it is important to note that a single-wire proportional counter with a drift radius of 5 cm, filled with an Ar-CO_2 gas mixture, can exhibit an ion drift time on the order of one millisecond

For timing measurements, the shaper bandwidth can be adjusted to produce pulse widths of approximately 0.1 μs. Operating with short signals enhances double-pulse resolution and is generally more convenient for both data acquisition and processing. However, as mentioned earlier, shortening the pulse can lead to significant pulse height fluctuations. The practical limit for pulse shortening depends on the available primary ionization (see Equation 1-3).

Another important parameter of a single-wire detector is its time resolution, which defines the precision with which the crossing time of a particle can be measured.

Time resolution depends on several factors, including:

- Detector geometry (electric field configuration)
- Primary ionization statistics
- Gas mixture and pressure
- Shaper bandwidth and electronics response
- Amplification and signal-to-noise ratio (SNR)

Typical single-wire proportional counters with a standard shaper achieve a time resolution of around 1 μs or slightly better. However, with carefully optimized parameters and electronics, the resolution can be improved to below 0.1 μs.

Let's now consider the signal fluctuations

In the simplest case of a point-like primary charge triggering the avalanche, the amplitude of the fast signal within a certain voltage range is proportional to n_0:

$$S_{fast}=kAn_0 \quad (2\text{-}4),$$

where k is a proportionality coefficient.

If monochromatic X-ray radiation is fully absorbed within the volume of a single-wire counter, the output signals from the shaper will not have a constant amplitude, as one might expect from an ideal detector. Instead, they will exhibit some variation in the amplitude distribution.

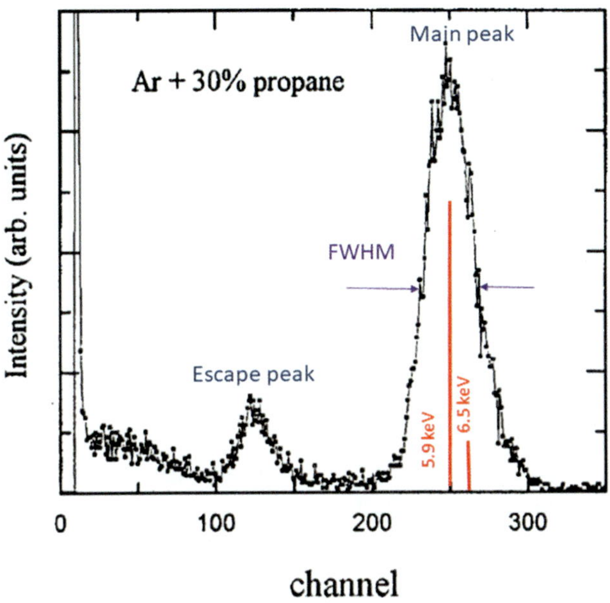

Figure 2-10. Pulse-height spectrum from a single-wire counter detecting X-ray radiation from a ^{55}Fe radioactive source FWHM stands for Full Width at Half Maximum (see explanation below). Adapted from [Broni, 2000, 6192-6200].

Fig. 2-10 shows a typical pulse-height spectrum of signals from a single-wire counter irradiated by a ^{55}Fe radioactive source. This source emits two X-ray lines: one at 5.9 keV and a less frequent one (10 times less) at 6.5 keV. The spectrum reveals a broad distribution of pulses from the shaper, with a peak cantered around channel #250, where both energy lines merge. Additionally, a smaller peak, known as the "escape peak," appears at channel #125. The origin of this peak will be discussed later.

Focusing on the peak at channel #250, three main factors contribute to its spread:

1. Primary Statistics – Fluctuations in the number of primary electrons (n_0) created by the absorption of monochromatic X-ray photons.

2. Avalanche Statistics – Variability in the multiplication of avalanche electrons.
3. Electronic Noise – Contributions from the charge-sensitive amplifier and the shaper.

A brief description of these factors follows.

1. In the case of independent ionizing collisions (Poisson statistics), the variance in the number of primary electrons is given by $\sigma=\sqrt{n_0}$. However, in ionizing processes, a key difference raises due to the constraint on the initial energy

$$n_0 \leq E_{abs}/W_i \qquad (2\text{-}5),$$

where E_{abs} is the energy absorbed in the detector volume and W_i is the mean energy required to create electron-ion pair respectively. This process deviates from Poissonian behaviour and exhibits reduced variance

$$\delta_{n0} = \sqrt{(n_0 F)} \qquad 2\text{-}6).$$

The factor F, commonly known as the Fano factor, typically ranges between 0.1 and 0.3 for most gases used in single-wire counters. Physically, F represents a degree of correlation between primary ionization events.

2. The avalanche process is governed by the probabilities of electron ionization collisions, making the situation complex and dependent on several parameters, including the number of primary electrons, gas gain, gas composition, pressure (p), and the diameter of the anode wire. For example, when an avalanche is triggered by a single electron, the typical pulse-height spectrum follows an exponential distribution. However, at a high E/p ratio, the single-electron avalanche distribution deviates from the exponential form, exhibiting a pronounced peak and reduced dispersion.

3. A widely used form of the gain distribution, P, was introduced in a previous study and is known as the Polya distribution (see, for example: [Breyer, 1973 91-93] and references therein)

$$P(S_{fas}, n_0) = (1/S_s)(S_{fast}/S_s)^{n0-1}[\{\exp(-S_{fast}/S_s)\}/(n_0-1)!] \qquad (2\text{-}7),$$

where S_{fast} represents the signal from the shaper, and S_s is the mean signal value. When $n_0 > 10$, the Polya distribution closely approximates a Gaussian distribution, making it possible to use standard statistical characteristics such as variance and relative variance.

According to Alkhazov [Alkhazov, 1969, 161-162], the relative gas gain variance is given by

$$(\delta A/A)^2 = f(A) \qquad (2\text{-}8)$$

The function f(A) increases with gain and at $A>10^3$ it reaches a plateau with a value between 0.7 and 1, depending on the gas composition

If one ignores the contribution of electronic noise and considers only the combined statistical effects of n_0 production and gas gain fluctuations, the energy resolution of an ideal detector can be expressed as

$$\delta S_{fast}/S_{fast} = \sqrt{[\{F+f(A)\}/n_0]} \qquad (2\text{-}9)$$

In most practical cases, however, electronic noise is not negligible. The energy resolution can then be estimated using the following formula:

$$\delta S_{fast}/S_{fast} = \sqrt{[[\{F+f(A)\}/n_0]+\delta_n^2\ ]]} \qquad (2\text{-}10).$$

After discussing the main factors influencing the energy resolution of a single-wire counter, let us revisit Fig. 2-10. The distribution below the peak at channel #250 can be characterized by the relative variance, δS_{fas} , or an equivalent parameter, the Full Width at Half Maximum (FWHM) of the spectrum curve, normalized by the amplitude. The relationship between these quantities is given by

$$FWHM = 2.35\ \delta S_{fast} \qquad (2\text{-}11).$$

Let us now briefly explain the origin of the peak at channel #125. The interaction of 5.9 keV and 6.5 keV X-rays with argon Ar atoms occurs through several channels. Photoionization can take place either from the outer shell or the first inner shell.

In the first case, the liberated photoelectron deposits all its kinetic energy, given by

$$E_k = E_v - E_i,$$

where E_v is the photon energy, and E_i is the ionization potential of the gas, within the detector volume. In this specific case, this process corresponds to a peak in the pulse-height distribution at channel #250.

However, photoionization is more likely to occur from the first inner shell. In this case, the ejected photoelectron will have a kinetic energy given by $E_k = E_v - E_{in}$, where E_{in} is the ionization energy of the inner shell. The resulting vacancy in the inner shell can then be filled through one of the following processes: either by the emission of an Auger electron, where the total deposited energy remains E_k, or by the emission of a fluorescent photon. If the latter escapes the detector volume, the deposited energy is reduced by the energy of this photon, leading to a peak at channel #125, known as the escape peak.

Another characteristic of a single-wire counter widely used in practice is its energy resolution, expressed as a percentage. It is defined as the standard deviation (σ) or the full width at half maximum (FWHM) of the pulse-height distribution produced by monochromatic X-ray radiation, divided by the mean pulse amplitude. In general, energy resolution characterizes the detector's ability to distinguish between two closely spaced energy lines or, more broadly, between radiations of similar energy. In avalanche mode operation, typically for $A \geq 10$ and up to some threshold $A < A_{crit}$, the signal amplitude S_{fast} exhibits an exponential dependence on the applied voltage. The exact behaviour is strongly influenced by the type of gas used in the counter, its pressure, and the detector's geometry, particularly the values of r_a and r_c.

To gain a deeper understanding of the underlying physics of a single-wire counter, it is useful to analyse the fundamental processes occurring in the detector under two extreme primary conditions:

1. Pure noble gases
2. Molecular gases

The operation of the detector in these two cases differs significantly and depends on several factors, particularly the gain versus voltage characteristics.

To simplify the analysis, these characteristics can be generalized: by adjusting and deforming the x-axis scale (voltages for each gas), the curves can be aligned so that their initial sections coincide. This approach offers a clearer depiction of the differences in detector behaviour.

In Fig. 2-11, such generalized gain curves and counting rates as a function of voltage for these two gases, pure noble and molecular, are presented.

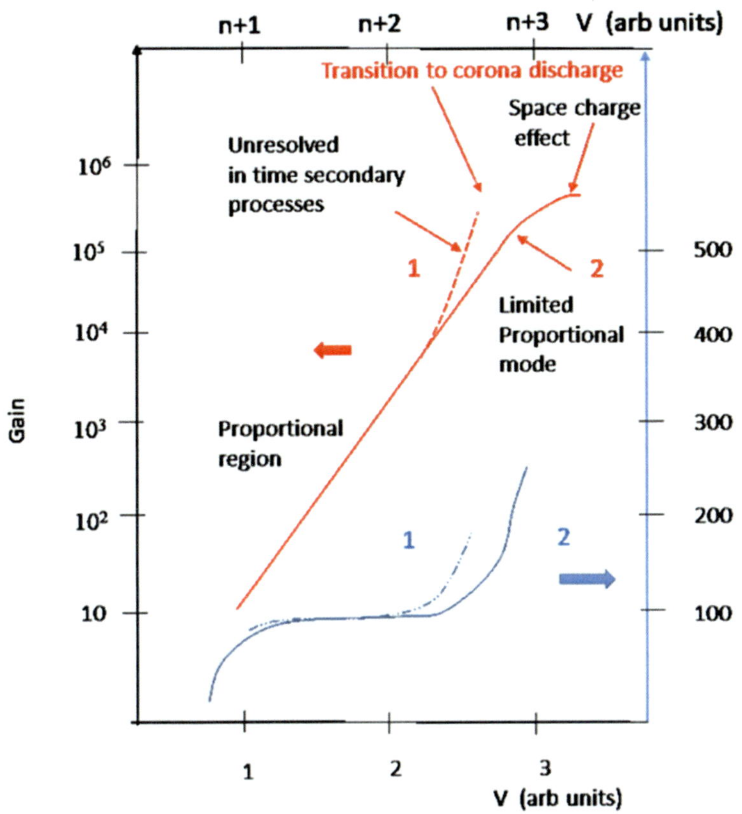

Figure 2-11. Generalized representation of gain (two upper curves) and counting rate (two lower curves) versus voltage for noble (1) and molecular (2) gases.

2.3. Operational features of single-wire Counters

Operation in pure noble gases

Looking at the schematically presented gain curve for the noble gas, one finds the following features:

a) At approximately A_f~5×10^3, the gain curve begins to increase more rapidly than an exponential function.

b) When the gain reaches around 10^5, the detector transitions into corona discharge.

Both of these critical values depend, of course, on the gas, its purity, and other conditions.

The commonly accepted explanation for the gain curve in noble gases increasing faster than exponentially when the gain exceeds a critical value A_{crit} is the occurrence of photon feedback pulses. Studies have shown that avalanches in noble gases emit strong excimer ultraviolet (UV) radiation [Peskov, 1979, 860-865], [Peskov, 1979, C7.331-C7 334]. Their spectra for Xe and Ar are depicted in Fig. 2-12. If the time resolution of the detector is insufficient to resolve individual feedback pulses, two or more pulses may be integrated by the electronics, merging into a single pulse with a larger amplitude. As a result, the pulse amplitudes begin growing faster than predicted by the exponential law.

The same figure also presents the counting rate versus voltage, assuming that 100 X-ray photons are absorbed by the detector each second. Initially, the counting rate increases, then reaches a distinct plateau, and finally rises sharply at the voltage where the gain curve deviates from its exponential behavior. One contributing factor to this is ion feedback; however, researchers have recently identified a more complex phenomenon, sporadic jets of electrons originating from the detector cathode [Iacobaeus, 2002, 1622-1628].

Let's now take a closer look at the physics behind the features formulated above.

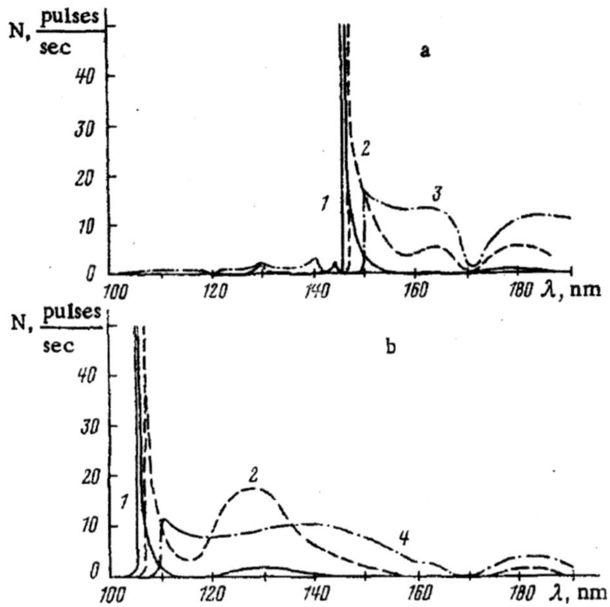

Figure 2-12. *Excimer emission spectra of Xe (a) and Ar (b) at various pressures: 1 – 0.1 atm, 2 – 1 atm, 3 – 10 atm, 4 – 25 atm. Reproduced from [Peskov, 1979, 860–865] with permission from Springer Nature.*

There are two main channels for excimer formation. The first occurs through collisions between excited atoms (A*) and surrounding atoms:

1. Electron impact excitation:

$$e_{hot}+A \rightarrow A^* + e_{th},$$

where e_{hot} and e_{th} represent electrons with high and low kinetic energy, respectively.

2. Excimer formation through atomic collisions:

$$A^*+A \rightarrow A_2^{*v}$$

$$A_2^{*v}+A \rightarrow A_2^*+A$$

Here, A_2^{*v} denotes excimers with vibrational excitation, while A_2^* represents excimers with pure electronic excitation.

3. Excimer dissociation and photon emission:

$$A_2^* \rightarrow A + A + h\nu$$

This process leads to the formation and radiative decay of excimers, emitting photons in the process.

The second channel for excimer formation begins with noble gas ions A^+, which undergo a series of transitions through collisions with other atoms and low-energy electrons:

1. Ionization by energetic electrons:

$$e_{hot} + A \rightarrow A^+ + 2e_{th}$$

2. Formation of molecular ions:

$$A^+ + A + A \rightarrow A_2^+ + A$$

3. Recombination with low-energy electrons:

$$e_{th} + A_2^+ \rightarrow A^{**} + A$$

Here, A^{**} represents an atom excited to upper energy levels.

4. De-excitation via atomic collisions:

$$A^{**} + A \rightarrow A^* + A + \text{heat}$$

$$A^* + 2A \rightarrow A_2^* + A + \text{heat}$$

5. Excimer dissociation and photon emission:

$$A_2^* \rightarrow A + A + h\nu,$$

This sequence results in excimer formation and subsequent photon emission during radiative decay.

Note that the latter channel is not very efficient in single avalanche mode, as electrons rapidly collect on the anode, leaving behind positive ions. This reduces the probability of their recombination with A_2^+ ions.

In pure noble gases, excimer radiation travels to the cathode with minimal absorption, generating free photoelectrons via the photoelectric effect. Due to the physics of this effect, quantum efficiency Q increases sharply with photon energy (see Fig. 2-13 for an example). In the far-ultraviolet region, where excimers emit, Q can reach high values, often approaching 1.

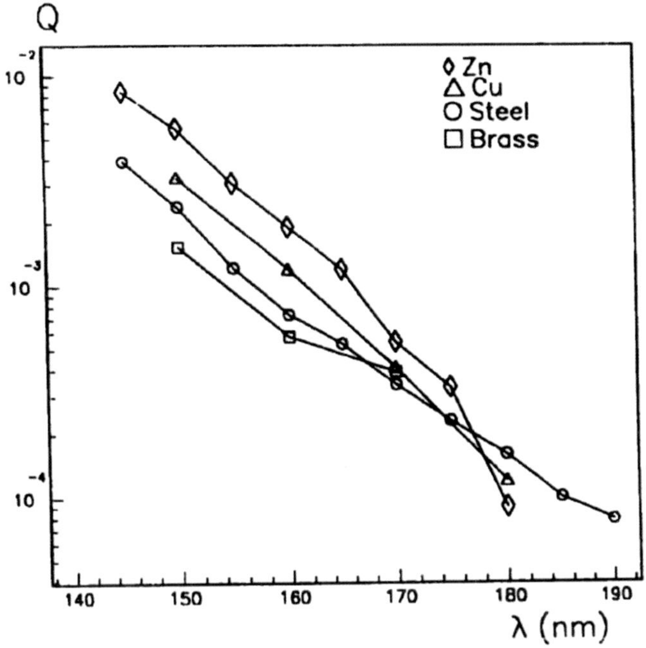

Figure 2-13. Typical dependence of metals' quantum efficiency on photon wavelength. Although the absolute values of quantum efficiency (Q) depend on specific conditions, such as surface quality, gas composition, and the electric field-to-pressure ratio (E/p), the overall trend remains consistent: Q increases sharply with photon energy. Reproduced from [Seguinot, 1990, 133-147] with permission from Elsevier.

The convolution of strong excimer emission with the high quantum efficiency of the metallic cathode results in a high value of the coefficient γ_{ph} , which determines the number of photoelectrons generated by the photoelectric effect per avalanche electron. For example, if an avalanche is initiated by a single primary electron and $A\gamma_{ph}=1$, it means that one secondary electron is produced per A avalanche electrons. If, at the same gain, the avalanche is initiated by n_0 primary electrons, then correspondingly, n_0 photoelectrons will be created.

Each secondary electron, during the drift time t_{dr}, moves toward the anode wire, where it triggers secondary avalanches. Typically, in a single-wire counter filled with Ar at 1 atm, $t_{dr} \approx$ a few microseconds. To achieve a high signal-to-noise ratio, the shaping time used in operation with a single-wire counter is often also a few microseconds. As a result, the primary and secondary pulses may overlap in time, as schematically shown in Fig. 2-14, producing a single pulse of larger amplitude. Due to this overlap, the mean signal value (dash curve in Fig. 2-11) begins to grow faster than the original exponential dependence

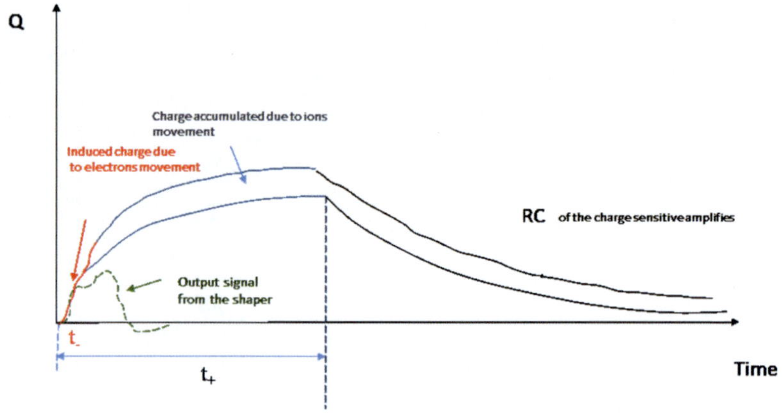

Figure 2-14. Overlapping in time of the primary pulse (lower curve) and smaller secondary pulses (not shown for simplicity), as observed at the output of the charge-sensitive amplifier (resulting pulse – upper curve). The shaped signal (dashed line) is also shown.

With modern electronics, resolving photon feedback pulses in time is not difficult. In this case, at voltages $V>V_{pf}$, the counting rate rises above the plateau level (see Fig. 2-11).

Similarly, a coefficient γ_{+}, can be introduced to represent the average number of secondary electrons generated per avalanche electron due to ion recombination on the cathode. This process occurs as follows.

After a primary avalanche happens and electrons are rapidly collected on the anode wire, the positive ions continue their slow movement toward the cathode and reach it after the time $t_{+} \gg t_{-}$.

When an ion comes extremely close to the cathode, within a fraction of a nanometer, an electron from the metal can tunnel through the lowered barrier into the ion's ground state, neutralizing it.

The excess energy

$$E_{virt} = E_i - \varphi \qquad (2\text{-}12),$$

(where φ is the metal's work function) is transferred via a virtual photon to another electron within the metal. If

$$E_{virt} > \varphi \qquad (2\text{-}13),$$

this electron can escape from the metal and enter the gas near the cathode. It then begins moving toward the anode wire, triggering a secondary avalanche that is delayed relative to the primary avalanche by approximately $t_{+}+t_{-} \approx t_{+}$.

As can be seen, the process by which a free electron appears near the cathode is somewhat similar to the photoelectric effect. Consequently, the efficiency of this phenomenon increases sharply with E_{virt}.

The ion feedback manifests as one or a series of well-separated pulses over time (see Fig. 2-15). This, in turn, affects the counting rate curve: even if the photon feedback pulses are unresolved in time, at certain voltages ($V>V_{if}$), the counting rate begins to rise above the plateau level.

In practice, in single-wire counters filled with noble gases, photon feedback dominates ion feedback. This is because, in noble gases, avalanches produce significantly more excimers and excited atoms than ions.

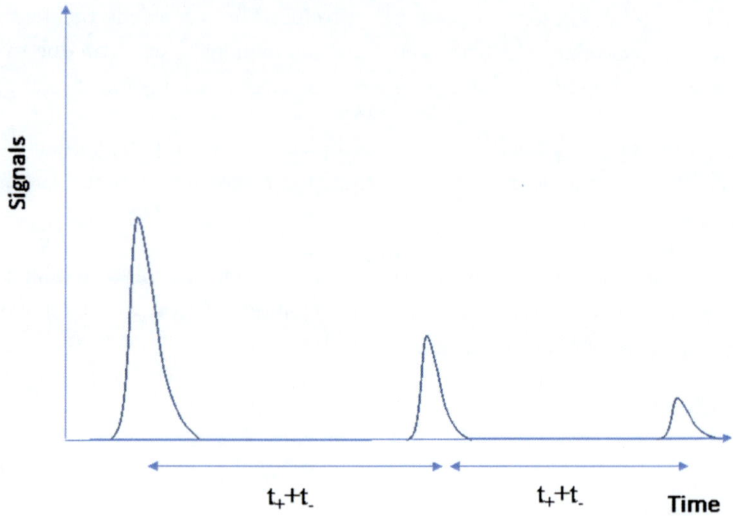

Figure 2-15. Schematic representation of ion feedback pulses detected using a low-time-resolution amplifier in the case where $\gamma_+ >> \gamma_{ph}$.

To explain this, we can refer to the well-known cross sections for excitation and ionization caused by electrons in noble gases. As an example, Fig. 2-16 shows the ionization cross section and the sum of excitation cross sections for argon as a function of electron kinetic energy E_k.

An electron drifting through a gas under the influence of an electric field gradually gains kinetic energy until it reaches $E_k>E_1^*$, where E_1^* is the excitation energy of the first electronic level of Ar. Upon undergoing an inelastic collision with an atom, the electron loses energy due to atomic excitation, and the process repeats.

In certain ranges of E/p, electrons will produce only excited atoms without causing ionization. This mode of operation is utilized in gas scintillation chambers. As the electric field increases beyond this regime, some electrons can acquire energy $E_k>E_i$, making ionization through electron-atom collisions possible.

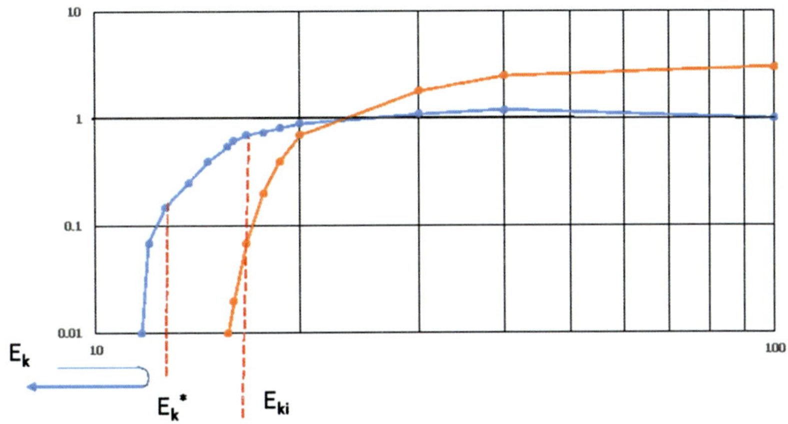

Figure 2-16. Total cross-section of excitation and ionization for Ar as a function of electron kinetic energy (adapted from https://magboltz.web.cern.ch/magboltz/).

However, even in this case, within certain energy intervals (for Ar, up to E_k<20 eV), the probability of excitation remains equal to or greater than the probability of ionization. This leads to the conclusion that avalanches in wire detectors filled with pure noble gases generate significantly more excited states (and consequently, excimers) than ions.

As a result, the photon-to-ion ratio (γ_{ph}/ γ_{+}) is typically much greater than one, with its exact value depending on the detector geometry and the E(r)/p ratio. Consequently, each ion feedback pulse will also include several photon feedback pulses [Peskov, 1976, 791-794], as illustrated in Fig. 2-17.

A mentioned before, the exact behavior of counting rate curve 1 in Fig. 2-11 depends on the time resolution of the amplifier used. Regardless, the counting rate increases sharply after a certain voltage threshold, ultimately leading to the onset of a corona discharge.

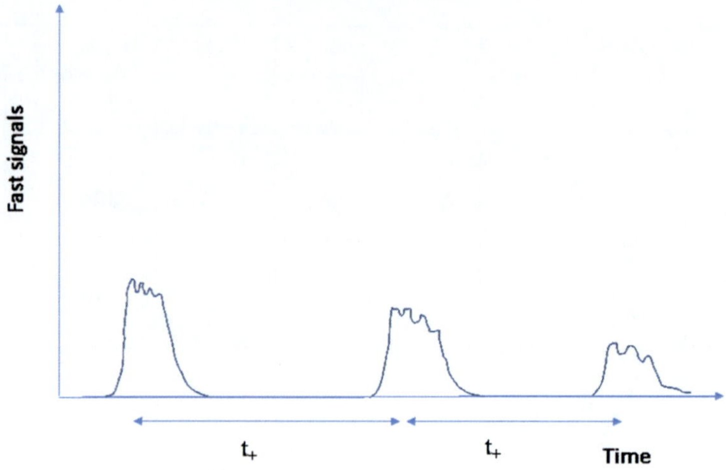

Figure 2.17. Schematic representation of feedback pulses detected using a high-time-resolution amplifier in the case where $\gamma_{ph} >> \gamma_+$.

The general conditions for the transition to corona discharge are:

$$A\gamma_+=1 \text{ or } A\gamma_{ph}=1, \qquad (2\text{-}14)$$

whichever occurs first.

As noted above, in pure noble gases, the condition $\gamma_{ph}=1$ is usually met first.

Benefits of operating single-wire detectors in molecular gases

In general, phenomena occurring in molecular gases are significantly more complex than those in pure noble gases. In most cases, their emission spectrum is spread over a wide spectral range, and UV emission is relatively weak, though there are exceptions, such as CF_4 and H_2.

For this reason, commercial UV lamps are typically filled with noble gases, mixtures of noble gases with H_2, or Hg vapours (D_2 or Hg).

As indirectly demonstrated in [Fonte, 1991, 140-145], in addition to molecular emission, UV photons from molecular dissociation fragments can also contribute to photoionization.

Another important characteristic of most molecular gases is their strong absorption in the UV region of the spectrum (see Fig. 2-18). As a result, the photoelectric effect is significantly suppressed.

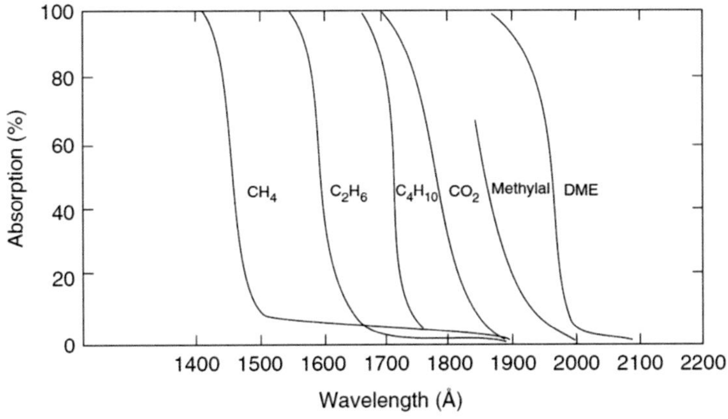

Figure 2-18. UV light absorption as a function of wavelength for several gases commonly used in gaseous detectors. The data correspond to a 5 cm gas thickness at 1 atm. Reproduced from [Fonte, 1991, 140-145] with permission from Elsevier.

Due to both weak emission and strong absorption of UV light, photon feedback is typically not observed in gases commonly used in single-wire counters. In these gases, γ_+ is also very small, allowing for extremely high gas gains, reaching up to 10^6–10^8.

Notably, at gains of $\geq 10^5$–10^6, detectors filled with molecular gases gradually transition from a proportional mode, where the signal amplitude is proportional to the number of primary electrons (see formula 2.4), to a limited proportional mode, where this relationship no longer holds. This transition is driven by the strong space charge effects in avalanches.

With a further increase in voltage, all output pulses become nearly identical, regardless of n_0. This marks a particularly interesting operational regime known as Geiger mode, in which the discharge spreads along the entire wire for a short duration, typically a few microseconds, before being intrinsically quenched due to the strong local space charge effect.

A schematic representation of this phenomenon is shown in Fig. 2-19. In this figure, r_{amp} represents the distance from the anode center where multiplication begins. Naturally, r_{amp} depends on the applied voltage across the detector: $r_{amp} = r_{amp}$ (V).

At sufficiently high charge levels generated by the avalanche, UV radiation from excited molecules and their fragments becomes intense enough to induce secondary avalanches through photoionization (Figs. 2-19a and 2-19b). These figures illustrate the moment when most of the electrons produced by the avalanche have been collected on the anode wire, while the more slowly drifting positive ions remain in their original positions within this timescale.

In Fig. 2-19, three key positions of photoelectrons, essential for further analysis, are shown.

A photoelectron in Position #1 (located above the primary avalanche) can generate only a relatively small secondary avalanche. This is because the electric field within the primary avalanche volume is significantly reduced due to the space charge. Consequently, the UV emission from this secondary avalanche is weak, producing only a negligible number of additional photoelectrons.

After one or two generations of this process, a dead zone forms within the primary avalanche, where the recreation of secondary electrons is suppressed. The radius of this zone is approximately equal to r_{amp}.

As a result, the zone where new avalanches can fully develop begins to propagate from the location of the primary avalanche in both directions along the anode wire (Figs 2-19b and 2-19c).

When the front of this moving plasma reaches the ends of the wire, the discharge is abruptly quenched due to the space charge. The current of the moving avalanches, I_G (typically ~1 μA), generates a voltage pulse across a resistor connected to the anode wire, given by $V = I_G R$.

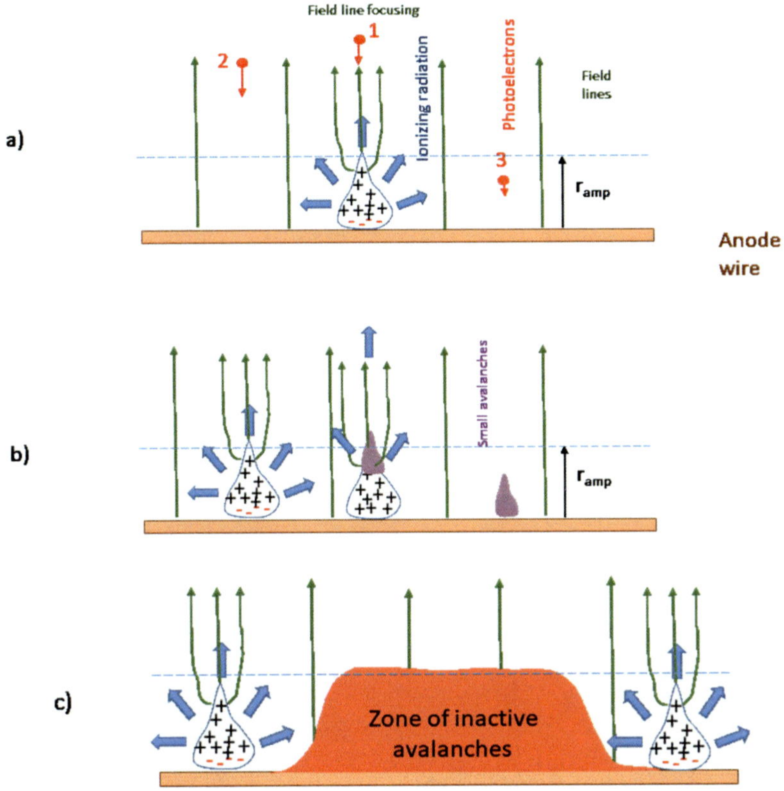

Figure 2-19. Schematic representation of the transition from the limited proportional mode to the Geiger regime:
(a) The primary avalanche generates photoelectrons.
(b) These photoelectrons trigger secondary avalanches.
(c) The strong space charge within each avalanche inhibits the formation of new avalanches in its vicinity, creating a zone free of additional avalanches that propagates along the wire.

These pulses can be safely observed directly on an oscilloscope using its 1 MΩ input, even without an amplifier.

In this mode of operation, all pulses have nearly identical amplitudes, regardless of the primary ionization value n_0 (see Fig. 2-20). Typically,

Geiger counters exhibit a wide counting rate plateau (see Fig. 2-11). Naturally, the amplitude of these pulses increases with the applied voltage. Figs 2-21 and 2-22 illustrate how the pulse amplitude evolves with increasing voltage.

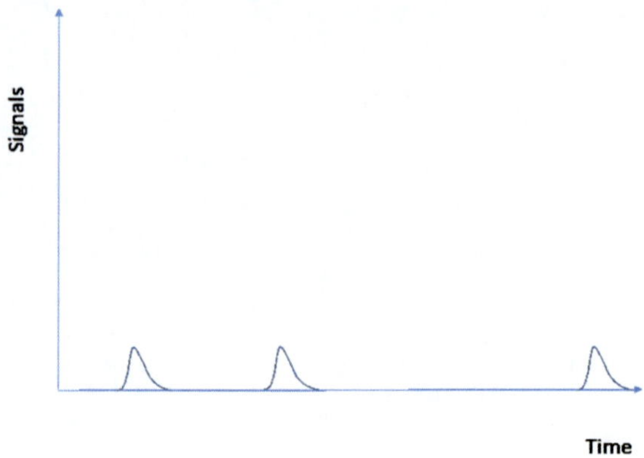

Figure 2-20. Schematic representation of pulses produced by external radiation in Geiger mode: they are practically equal in amplitude, which increases with voltage but does not depend on the value of primary ionization.

For instance, Fig. 2-21 depicts a scenario where $A\gamma_+$ approaches unity, causing ion-induced feedback pulses to accompany almost every primary pulse. As a result, the counting rate increases sharply (see curve 2 in Fig. 2-11 at V>2.3 in arbitrary units).

Fig. 2-22 illustrates the case where $A\gamma_+ \sim 1$. At this increased voltage, some pulses begin to trigger short bursts of corona discharge. With a further increase in voltage, a stable corona discharge emerges.

The corona current may exhibit slight, nearly periodic oscillations (see Fig. 2-22). This behaviour is a form of ionization instability caused by the corona space charge (see [Peskov, 1975, 1588-1590]).

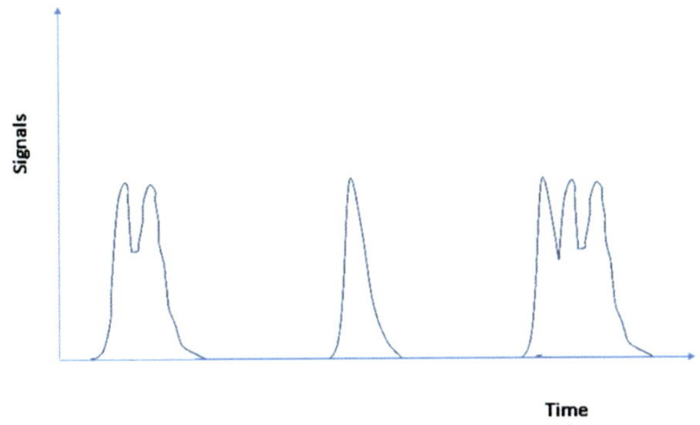

Figure 2-21. As the voltage increases, secondary pulses caused by ion feedback appear.

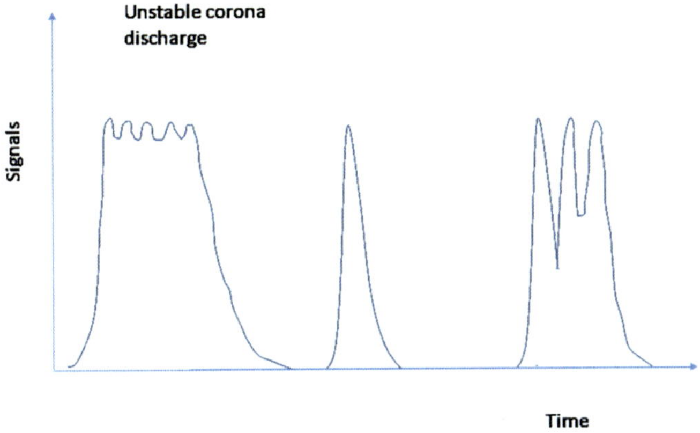

Figure 2-22. With a further increase in voltage, some radiation-induced pulses begin to trigger an unstable corona discharge.

Operation of Single-Wire Counters in Mixtures of Noble and Molecular Gases

In practice, most single-wire counters are filled with mixtures of noble gases and one or more molecular gases. This is done for two main reasons:

1. The operational voltage in such mixtures is generally lower than in pure molecular gases.
2. The addition of noble gases increases the stopping power for detecting radiation compared to most molecular gases.

The excitation and ionization processes in these mixtures are significantly more complex than described above. A comprehensive review of these processes would require extensive discussion and is beyond the scope of this book. Therefore, we will focus on the simplest case: a detector filled with a binary mixture of a noble gas and a sufficiently high percentage of a molecular gas, where $E_{in}>E_{im}$, and E_{in} and E_{im} represent the ionization potentials of the noble and molecular gases, respectively.

In such a mixture, the final stage of transition chains involves all noble gas ions and excited noble gas atoms, produced by the avalanche, transferring their energy to surrounding molecules. Consequently, only molecular ions or dissociation fragments reach the cathode, resulting in a low value of γ_+.

The emission of UV light from the avalanche is also primarily governed by excited molecules. This is because noble gas excitation and ionization by electron impacts are significantly suppressed, as described in Section 2.4 and illustrated in Fig. 2-16. Moreover, any UV photons emitted by the noble gas will be strongly absorbed by the molecular components of the mixture. As a result, γ_{ph} remains as low as in a pure molecular gas.

2.4. Parallel-plate avalanche chamber

Townsend conducted his avalanche studies using a parallel-plate electrode geometry. It is not surprising, therefore, that some of his successors adopted the same configuration for a more detailed investigation of specific processes occurring during avalanche development.

This choice was well justified, as the parallel-plate geometry was particularly convenient for visualizing avalanches using the Wilson cloud chamber (see Fig. 2-23).

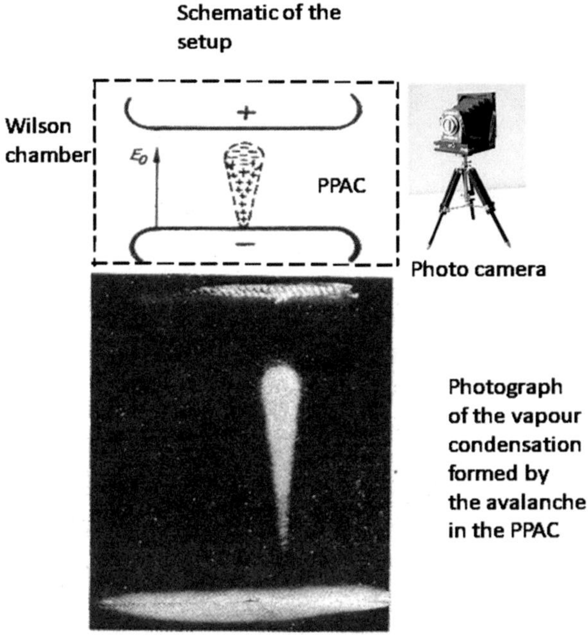

Figure 2-23. Photograph of a Townsend avalanche in a parallel-plate avalanche chamber placed inside a Wilson chamber. Adapted from [Raether, 1964, 1-191].

Of course, the most practically significant results were obtained using single-wire counters (some of which were discussed in previous paragraphs). However, the parallel-plate geometry provides a simpler interpretation of the results and facilitates easier calculations of key parameters, such as avalanche current and induced signals.

Let's illustrate this with several examples.

The details of avalanche development, associated processes, and the characteristics of current signals induced on the electrodes of a parallel-plate chamber (shown in Fig. 2-1) have been extensively studied in numerous works. One of the most comprehensive overviews can be found in [Raether, 1964, 1-191].

If n_0 electrons are created in a parallel-plate chamber with gap d and an electric field E is applied, these electrons will begin moving toward the anode with a constant velocity $v_-(E)$. The resulting current on the electrodes (the collect charge on the electrode divided by the collection time) is given by:

$$I_-(t) = en_0v_-(E)/d \quad (2\text{-}15),$$

where e is the elementary charge of an electron. For simplicity, diffusion effects are neglected in this consideration.

Similarly, if n_0 positive ions (instead of electrons) are generated, their motion will produce a current given by:

$$I_+(t) = en_0v_+(E)/d \quad (2\text{-}16),$$

where v_+ is the drift velocity of the ions.

Now, let's consider the case where n_0 primary electrons and ions are simultaneously created at a distance x from the anode. In this scenario, the total current induced on the detector electrodes will be the sum of the currents produced by both electrons and ions, as schematically illustrated in Fig. 2-24.

Note that $v_- >> v_+$, typically by 2–3 orders of magnitude. However, for simplicity, Fig. 2-24 illustrates a case where $v_-=10V_+$.

Within certain ranges of electric field values (which may differ for electrons and ions), there exists a simple relationship between the electric field E and the drift velocities:

$$v_- = \mu_- E \quad (2\text{-}17)$$

$$v_+ = \mu_+ E \quad (2\text{-}18)$$

where μ_- and μ_+ are the mobility coefficients for electrons and ions, respectively.

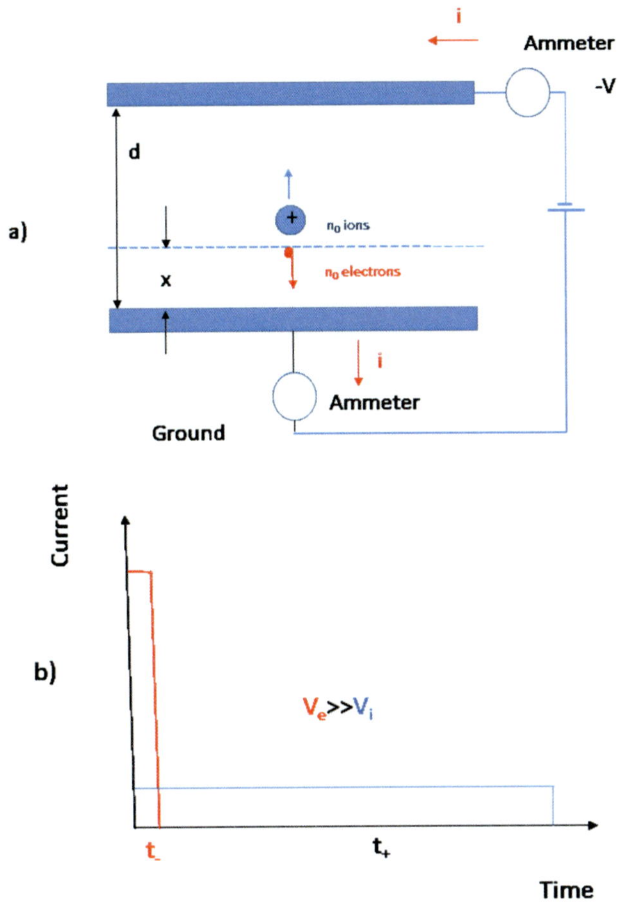

Figure 2-24. Analysis of the current flow in a parallel-plate chamber operating at a relatively low electric field in charge collection mode:
a) Schematic of the detector
b) Electron and ion currents as a function of time for a hypothetical case where $v_- \approx 10v_+$

In practice, charge-sensitive amplifiers are often used in radiation measurements (see Fig. 2-25a) because, as mentioned earlier, they offer significantly higher sensitivity than picoammeters. These amplifiers integrate the current over a specific time interval (typically longer than t_+). The output signal from such an amplifier is schematically illustrated in Fig. 2-25b.

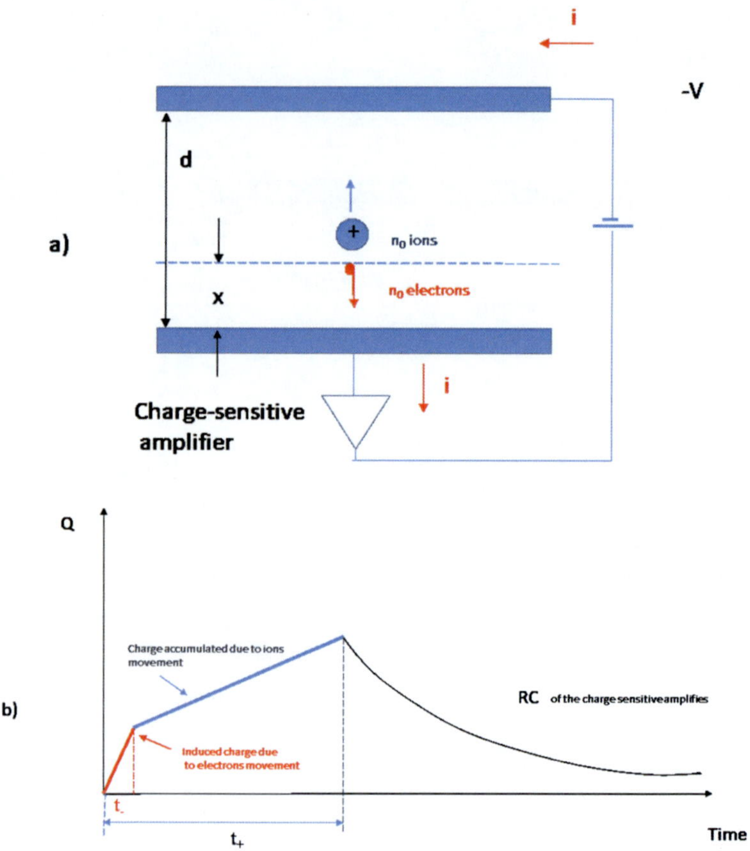

Figure 2-25: a) Schematic of a parallel-plate detector read out by a charge-sensitive amplifier; b) Signal from a charge-sensitive amplifier for the case presented in Fig. 2-24, showing two distinct components corresponding to electron and ion collection.

The charge-induced signals are proportional to the voltage drop experienced by the charge as it moves within the detector. This is a general principle, and in the specific case of a parallel-plate chamber, it becomes particularly evident:

$$Q_{el}/Q_{ion} = x/(d\text{-}x) \qquad (2\text{-}19)$$

where x is the distance from the anode at which the electron-ion pair is created.

For constant v_- and v_+ (which holds in a uniform electric field E), the induced charges are given by:

$q_-=i_-t_-$ and $q_+=I_+t_+$.These charges result from the motion of electrons and ions, respectively. The times t_- and t_+ are proportional to the drift distances of electrons (x, see Fig. 2-25a) and ions (d-x), which, in turn, correspond to the voltage drop experienced by the respective charges.

Now, let's consider a scenario where n_0 primary electrons, created at a distance x from the anode, move in a sufficiently strong electric field, allowing the multiplication process to occur. In this case, the current is given by:

$$I_-(t) = en(t)v_-/x \qquad (2\text{-}20)$$

where the number of electrons at time t follows:

$$n(t) = n_0\exp(\alpha v_- t) \qquad (2\text{-}21)$$

Substituting equation (2.21) into (2.20), we obtain:

$$I_-(t) = [en_0\exp(\alpha v_- t)v_-]/d \qquad (2\text{-}22)$$

As seen, the electron current increases exponentially until all electrons are collected at the anode (assuming diffusion is neglected), see Fig. 2-26. The ion current will increase exponentially over time t_-. However, similar to the previously considered case, it will remain constant during the interval $t_- > t < t_+$, as it is generated by a cloud of positive ions moving at a constant velocity $v_+= \mu_+E$. Since this ion cloud is not point-like, as assumed in Fig. 2-26, but has a finite size, the current will gradually diminish toward the end of the collection period.

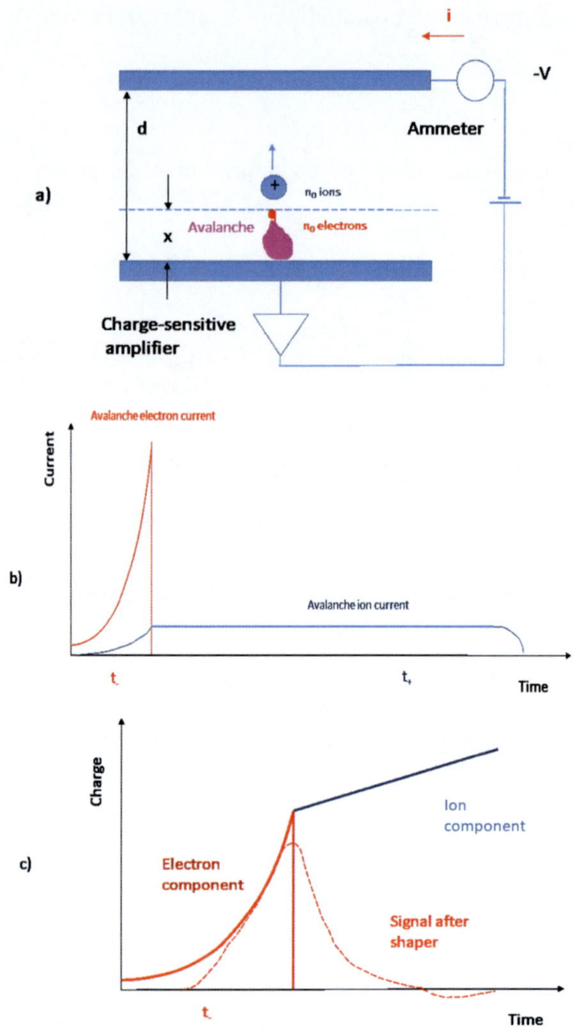

Figure 2-26. Schematic representation of signals produced by an avalanche in the PPAC:
(a) Diagram illustrating current and charge measurements.
(b) Current signal.
(c) Signal from the charge-sensitive amplifier.

We do not provide an analytical formula for the current produced by positive ions over time, as this slow component of the signal is generally insignificant in practice. In most cases, only the fast component is considered.

In the general case, the amplitude of the fast signal after the charge-sensitive amplifier (for a point-like primary charge n_0 is given by:

$$S = k\ (n_0,\ x)\ n_0 \exp \alpha x \qquad (2\text{-}23),$$

where x is the distance of the primary charge from the anode plate, and k is a coefficient.

Since the signal amplitude depends not only on n_0 but also exponentially on *x*, the PPAC is not an ideal device for measuring n_0.

At high electric fields, avalanches triggered by passing particles can transition into sparks. By photographing these sparks, researchers were able to roughly determine the position of the particle track traversing the detector (Keuffle, 1949, 202-208], [Bella, 1953 1338-1340] and [Bella, 1953, 1461-1479]. Using a stack of such detectors (see Fig. 2-27), it was possible to obtain 2D and even 3D images of particle tracks.

Initially, a constant voltage difference was applied between the plates (electrodes), but this often led to spontaneous discharges, causing unstable operation. However, with the use of scintillator counters, it became possible to apply the potential difference only at the moment a particle passed through the plates, improving stability and reliability.

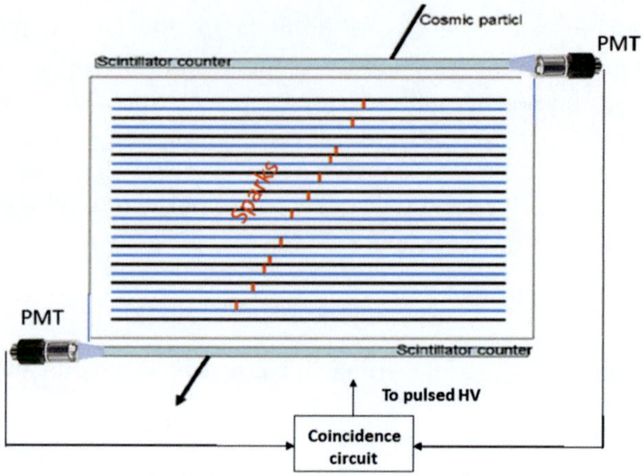

Figure 2-27. A schematic representation of a stack of spark counters equipped with two scintillators. When a particle passes through, it generates a light pulse in the scintillator, which serves as a trigger for applying a short high-voltage pulse to the spark counters. At this moment, sparks are produced in each parallel-plate structure, making it possible to visualize the particle's track. (Adapted from https://www.hep.phy.cam.ac.uk/~lester/teaching/SparkChamber/NIKHEF/ eppog.web.cern.ch/eppog/Resources/SparkChamber.html).

Concluding Remarks

The single-wire counter described in this section served as the foundation for Charpak's invention of the multi-wire proportional chamber, which revolutionized the field of charged particle and photon detection.

Charpak also had experience with spark chambers, some of which featured mesh electrodes that allowed sparks to be photographed in three dimensions. This likely inspired him to modify the classical parallel-plate avalanche chamber (PPAC) design by replacing the solid electrodes with meshes and introducing an additional gap. This seemingly minor modification transformed an academic device into a practical detector. These mesh PPACs will be discussed in the next chapter.

CHAPTER 3

THE MWPC AND MESH PPAC

"All truth is simple. All falsehood is sophisticated."
Hryhorii Skovoroda, Ukrainian philosopher

Abstract

This chapter presents classical designs of multi-wire proportional chambers (MWPC) along with the underlying physics principles governing their operation. Additionally, the transformation of a parallel plate avalanche chamber (PPAC) into a practical device by Georges Charpak is discussed.

3.1. MWPC, principle of operation, and main characteristics

Genius ideas are often remarkably simple. Charpak's concept was no exception, what once seemed revolutionary may now appear trivial. However, many groundbreaking ideas follow this pattern: they seem complex at first but become elegantly simple in hindsight.

Consider, for example, the photoelectric equation proposed by A. Einstein in 1905:

$$hv = \varphi + E_k \quad (3\text{-}1),$$

where hv is the energy of an incident photon, ϕ is the metal's work function, and E_k is the kinetic energy of the emitted photoelectron.

Today, even students recognize that this formula simply expresses the law of energy conservation: the energy of an incident photon is used to overcome the metal's work function, with any excess energy manifesting as the kinetic energy of the emitted electron. However, the true genius behind this idea lies in the concept that light exists in discrete packets, photons, with energy hv, which can be absorbed by individual electrons. Despite its simplicity, the profound implications of this equation earned Einstein the Nobel Prize in Physics in 1921. Now, let's return to Charpak's invention. What was the key idea behind it?

It was remarkably simple: instead of using a single-wire counter for particle detection, he proposed an array of them. By placing parallel anode wires between parallel cathode plates and applying a voltage, a well-defined pattern of equipotential lines is formed, as illustrated in Fig. 3-1.

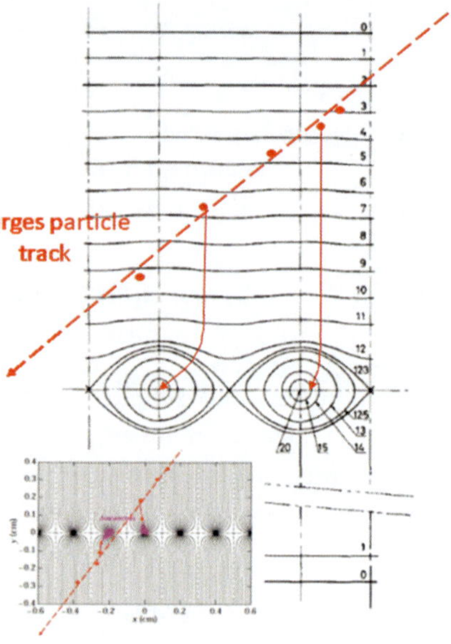

Figure 3-1. The map of equipotential lines in a multiwire detector. The detector consists of wires with a diameter of 4 × 10^{-3} cm and a pitch of 0.3 cm. A potential difference of 20 V is applied between the wires and the external mesh (adapted from [Charpak, 1968, 262-268]).

As observed in the close vicinity of the anode wires, they exhibit circular shapes similar to those in a conventional single-wire counter. The key innovation of this design is the absence of separating walls between the wires, making the detector both simpler and more compact.

So, what happens when a charged particle traverses such a detector? Typically, it creates a track consisting of electron-ion pairs. Under the influence of the applied electric field, the primary electrons move toward the anode wires, while the ions migrate toward the cathode electrode. In general, they follow the field lines while also undergoing diffusion. As a result, in the first approximation, electrons drift almost perpendicular to the equipotential surfaces, with each eventually approaching a corresponding wire (see Fig. 3-1).

The current induced by the motion of ions remains very small until the electrons reach the vicinity of the anode wires. At sufficiently high voltages applied to the detector electrodes, the electrons initiate Townsend avalanches. At 1 atm, these avalanches typically begin developing very close to the anode wire, at a distance roughly between three and five times the anode radius.

Accordingly, the gas gain can be calculated using the following formula, which is analogous to Eq. 2-2:

$$A(r) = \exp \int_{ra}^{rm} \alpha(r)dr = \exp \int_{ra}^{5ra} \alpha(r)dr \qquad (3\text{-}2).$$

As an example, in Fig. 3-2, the gain curves as a function of applied voltage, measured with an MWPC operated in various gas mixtures, are shown. By comparing Figs 3-2 and 2-11, one can conclude that the curves exhibit the same characteristics: the gas gain increases exponentially with the applied voltage, and at gains close to 10^5, saturation begins to occur due to the space charge effect.

In the earliest designs of MWPCs, each wire was connected to its own amplifier. According to classical electrodynamics, the signal induced on the wire closest to the avalanche is negative, while the signals on neighbouring wires are positive. By using amplifiers sensitive to negative signals, it is possible to determine the wire near which the avalanche occurred.

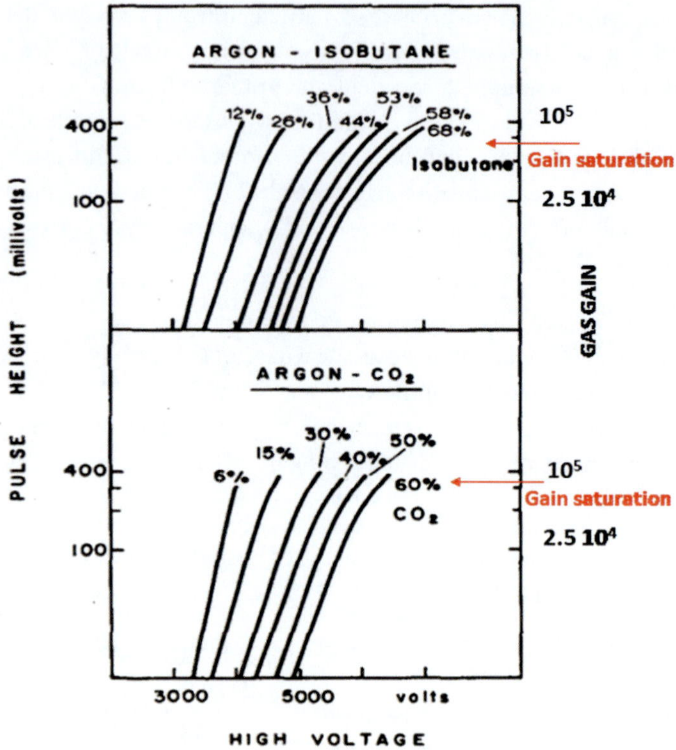

Figure 3-2. Gas gain curves of an MWPC operating in various gas mixtures at a total pressure of 1 atm. A signal amplitude of approximately 400 mV corresponds to a gain of 10^5(adapted from [Bouclier, 1970, 149-161]).

When a single ionization event is recorded, its position can be determined, in a simple approximation, with an accuracy equal to the wire pitch. However, if all that is known is that the particle passed somewhere within one wire pitch d, and if it can be assumed to be equally likely to be anywhere in that interval, then the uncertainty is given by the standard deviation of a uniform distribution. As a result, the achieved position resolution will be several times better than the wire pitch. Let's break this down simply.

In general, one of the key properties of a detector is its resolution for measuring a given quantity Z. If z_r represents the detector's response to this quantity, the resolution is defined as the standard deviation σ_z (or alternatively, the full width at half maximum, FWHM) of the distribution $D(z_r)$ of the measured values z_r, assuming a delta function input distribution. This is why the position resolution will be better

For example, as mentioned earlier in Chapter 2, in the case of monochromatic X-rays, when a sufficiently large number of primary electrons are created by each absorbed photon, the response distribution can be approximated by a Gaussian, where the FWHM is equal to 2.36 σ_z.

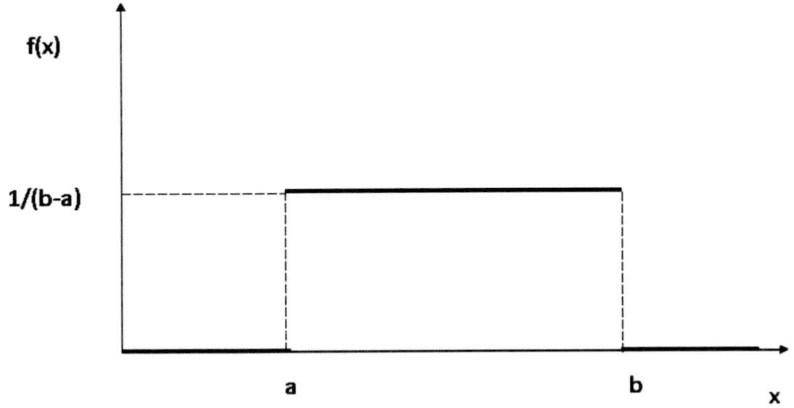

Figure 3-3. Probability density function for the MWPC exposed to a uniform flux of ionizing radiation (the area under the curve is equal to one).

For a uniform spatial distribution, such as an MWPC irradiated by evenly distributed radiation, the density function follows a square distribution (see Fig. 3-3) and is given by:

$$f(x) = 1/(b-a) \tag{3-3}$$

(for $a \leq x \leq b$ and 0 otherwise), where a and b are the coordinates defining the boundaries of a given detector "cell," with a wire positioned at the center.

Let $E_p(x)$ be the expected value of the random variable X. The mean of the uniform distribution is given by:

$$m_x = E_p(x) {}_a\int^b x/(b-a)\, dx = (b^2-a^2)/2(b-a) = (a+b)/2 \qquad (3\text{-}4)$$

One can also calculate:

$$E_p(x^2) = {}_a\int^b x^2 dx\, /(b-a) = (a^2+ab+b^2)/3 \qquad (3\text{-}5)$$

The variance is then given by:

$$\sigma^2 = E_p[(x - m_x{}^2)] = E_p(x^2) - m_x{}^2 = [(a^2+ab+b^2)/3] - [(a+b)^2/2] = (b-a)^2/12 \qquad (3\text{-}6)$$

Therefore, the standard deviation, which is the square root of the variance, is:

$$\sigma = (b-a)/\sqrt{12} \qquad (3\text{-}7)$$

As follows from the consideration, the position resolution of MWPC is limited by

$$\sigma_x = d/\sqrt{12}, \qquad (3\text{-}8)$$

where d is the wire pitch or the size of the individual "cells" containing an anode wire. Note that this formula represents a *worst-case* scenario: a geometry-only resolution corresponding to a so-called binary readout. Any real MWPC that exploits analogue information or charge sharing can (and usually does) surpass this limit. Soon after, Charpak and his collaborators developed a modified MWPC design in which the cathode planes were also made of wires (see, for example, [Charpak, 1971, 377-388]).

This not only reduced the material budget but also allowed amplifiers to be connected to the cathode wires, enabling the recording of signals induced by ion motion. Recall that electrons and positive ions produced in an avalanche induce signals as they move toward the electrodes: a negative signal on the wire where the avalanche occurs and positive signals on all surrounding wires. To minimize the number of electronic channels, wires are often grouped into strips. The profile of the positive induced signals on cathode wires or strips is quite broad, with FWHM of approximately 2h, where h is the anode-cathode distance. However, the centroid of this distribution is well defined (see Fig. 3-4). By determining its maximum,

X_{center}, one can achieve a position resolution along the anode wire between 50 and 300 μm, depending on the readout methods and electronics used.

One of the most common methods for this determination is the "centre of gravity" method:

$$X_{center} = \sum x_i S_i / \sum S_i, \qquad (3\text{-}9)$$

where x_i and S_i are the central coordinates and the signal recorded on strip i, respectively (see Fig. 3-4).

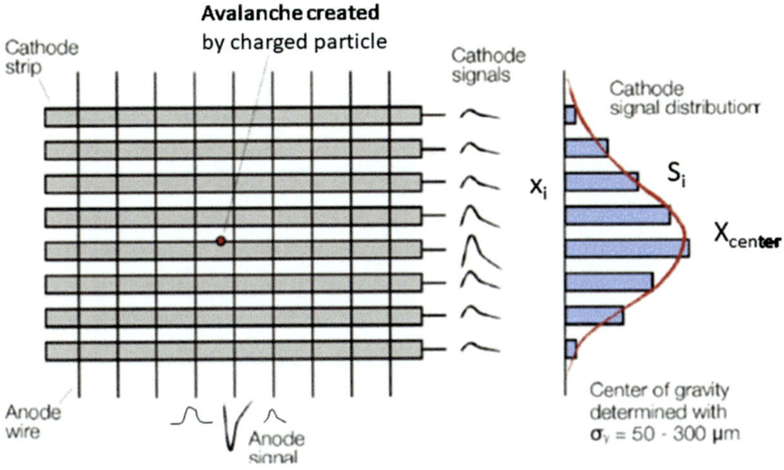

Figure 3-4. Schematic representation of induced signals on cathode strips and their distribution in the case of a single avalanche (adapted from http://www.desy.de/~garutti/LECTURES/ParticleDetectorSS12/L4_gasDetectors.pdf).

Signals induced on wires/ strips can be calculated analytically using Ramo theorem [Blum, 2008,157-180]:

$$I\,(I, \lambda_s, w) = I_c\,(t) P_0(\lambda_s) \qquad (3\text{-}10),$$

where λ_s=x/h –the of the distance of the cathode strip from the avalanche position and $P_0(\lambda_s)$, is a pad response function, which determines what fraction of the total cathode signal is induced in the strip as a function of the distance between the center of the strip and the avalanche position measured in the cathode plane. It has a quite complicated formula (see [Blum, 2008,157-180]). The fact that the signal on the entire cathode plane is equal to half of the total cathode signal can be verified mathematically, by finding the limit $P_0(\lambda_s)$ when w is increased to infinity. We do not present here details, but analytical calculations fully confirm the expected result.

Examples of the calculation of $P_0(\lambda_s)$ are shown in Fig. 3-5. As can be seen, as w decreases, the curves become smaller and thinner.

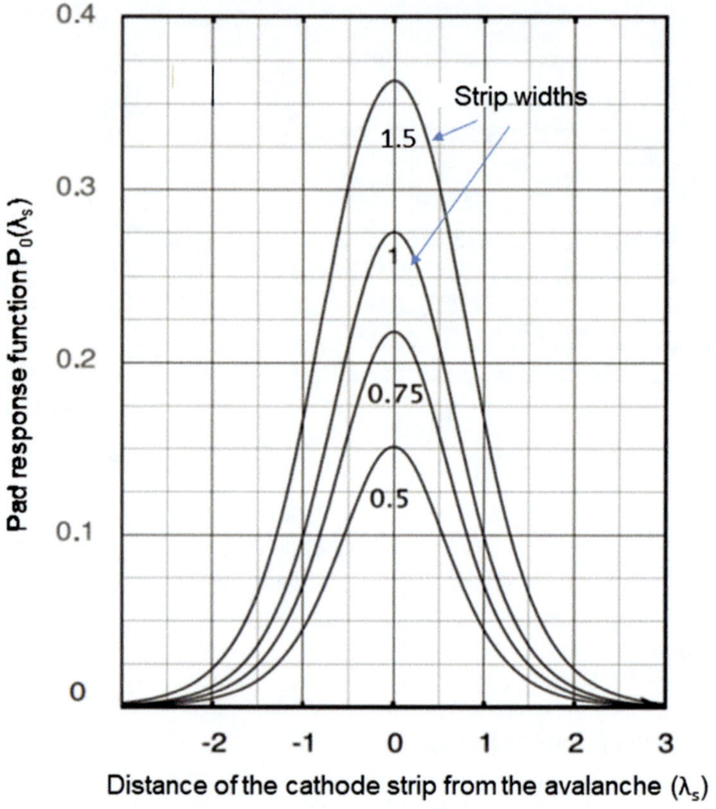

Figure 3-5. Pad response function versus λ_s for four different strip widths w, expressed in units of the anode-cathode distance h. As w decreases, the curves become smaller and thinner (adapted from [Blum, 2008, 157-180]).

In the case of a track, a more complex effective pad function, P_{eff} , can be introduced to properly account for the distribution of avalanche positions that contribute to the cathode strip signal (Fig. 3-6)

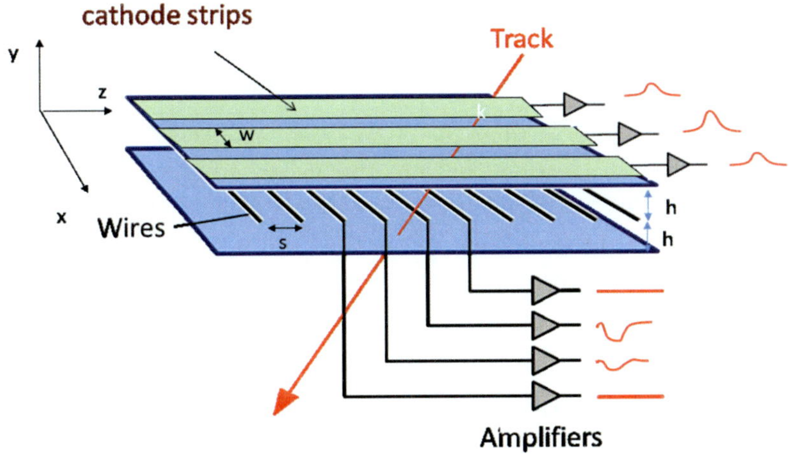

Figure 3-6. Schematic representation of induced signals in the case of an inclined track creating two avalanches. Signal amplitudes and their distribution are used to calculate the induced signal profile and, if necessary, determine its center of gravity (adapted from wikepedia?https://en.wikipedia.org/wiki/Wire_chamber, licensed under CC BY-SA 3.0.).

In some MWPC designs, one of the cathode planes may deviate from the traditional wire or strip structures. For instance, it can consist of metallic pads that are electrically isolated from one another (see description of ALICE Cherenkov detector in Chapter 7).

Calculating the induced signals on various cathode structures is beyond the scope of this book. However, it is worth noting that cathode strip readout enables a spatial resolution of approximately 100 μm for determining the avalanche position along the anode wire. While the exact resolution depends on multiple factors, it generally remains within the same order of magnitude.

The time resolution of a MWPC is influenced by the jitter in the position of primary electron clusters, particularly those near the anode wire, created by ionizing radiation within the drift region. Typically, the resolution is around 0.1 μs or slightly better, depending on factors such as the gas mixture used, gas gain, electronics, and the data analysis algorithm. For example, a study [Patra Rajendra Nath, 2018, 509 - 512] reported a time resolution of approximately 10 ns (sigma) after optimal parameter adjustments.

Energy resolution is another key characteristic of MWPCs, determining how precisely the detector measures the energy of absorbed radiation. Ideally, for a given energy deposition in the gas, the resulting pulse-height readout would form a delta function. However, in reality, energy measurement is influenced by various processes, such as gas ionization, electron multiplication, and signal readout, that introduce fluctuations in pulse height (see formulas 2-9 and 2-10).

Fig. 3-7 presents a typical pulse-height spectrum measured with a ^{55}Fe source. It closely resembles the one shown in Fig. 2-10, which is expected since an MWPC can be considered an array of single-wire counters. The typical energy resolution of an MWPC, depending on its geometry, gas gain, composition, and electronics, ranges from 15% to 17% FWHM.

Finally, let's examine the rate capability of the MWPC, the maximum number of events per second that the detector can register without significant information loss due to space charge effects and signal pile-up.

The space charge effect arises in high-rate environments due to the accumulation of positive ions in the detector volume, which distorts the electric field, reducing gain and efficiency. This effect depends on factors such as gas gain, ion mobility, and the size of the drift region. In MWPCs, the drift region is considerably smaller than in ordinary single-wire counters, resulting in a correspondingly higher rate capability.

Signal pile-up, on the other hand, occurs when signals from closely spaced events overlap, potentially leading to misinterpretation or loss of information.

These and other issues have been addressed in several studies. For example, in [Hall, 1995, 242-245], an MWPC with a very small distance (0.3 mm) between the anode wire plane and the cathode was investigated.

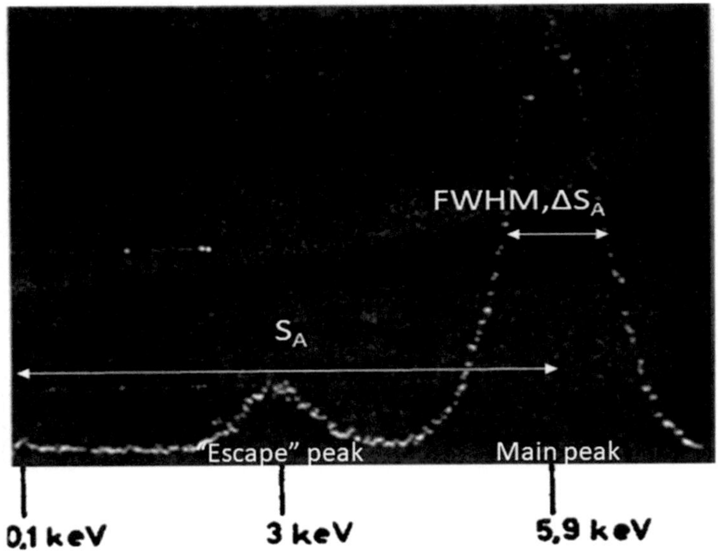

Figure 3-7. Typical pulse-height spectrum measured with an MWPC irradiated by a ^{55}Fe source, which emits two X-ray lines at energies of 5.9 keV (strong) and 6.5 keV (weak). The energy resolution is defined as FWHM=$\Delta S_A/S_A$ and, in this case, is 15%. The peak at 3 keV is a so-called "escape peak," which appears when the chamber is filled with a gas mixture containing argon (for details, see Chapter 2). Adapted from [Charpak, 1968, 262–268].

Fig. 3-8 presents the rate characteristics of both a microgap MWPC and a conventional MWPC. The intensity of the X-ray beam was controlled using an attenuator placed in front of these chambers, which can be described by the following equation:

$$N_{out} = N_{in}\, Exp(-k_a x_a) \qquad (3\text{-}11)$$

where N_{in} is the number of photons per second entering the absorber, k_a is the absorption coefficient of the attenuator, x_a is its thickness, and N_{out} is the number of photons per second passing through the absorber.

The number of detected photons is given by:

$$N_{dph} = K_{eff} N_{out} \quad (3\text{-}12),$$

where K_{ef} is the detection efficiency of the chamber.

As seen from the curves in Fig. 3-8, when N_{dph} >3000 ph/mm²/s, the detection efficiency of a conventional MWPC begins to drop dramatically, whereas in the case of a microgap detector, it remains constant. This decline is primarily caused by two factors: the space charge effect from accumulated positive ions and signal overlap in time.

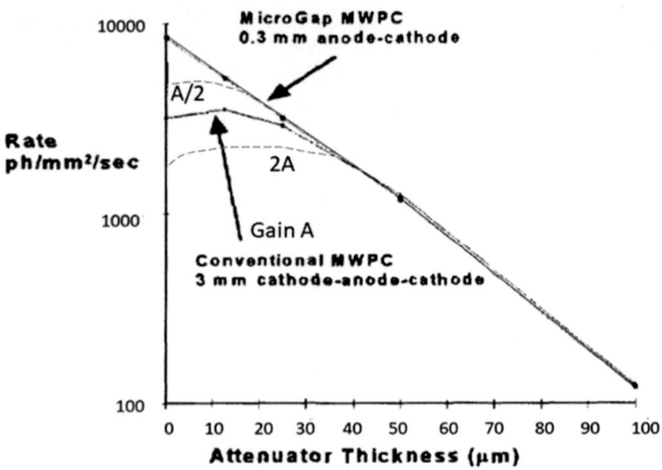

Figure 3-8. Number of detected X-ray photons as a function of attenuator thickness, measured using a conventional MWPC and a microgap MWPC. Dotted lines indicate the expected behaviour at gas gains of A/2 *and* 2A *(adapted from [Hall, 1995, 242–245]).*

The space charge effect reduces gas gain, resulting in lower signal amplitudes. As a result, fewer pulses exceed the electronic threshold and are recorded. Additionally, the pile-up effect causes overlapping pulses to be counted as a single event by the electronics, further decreasing the number of registered pulses.

As shown in Fig. 3-8, reducing the gap between the anode and cathode structures is a highly effective method for achieving high-rate capabilities. This approach serves as the foundation for modern micro-pattern detectors. For more information, see [Francke, 2014,1-303].

3.2. Mesh PPAC

In the early 1960s, Georges Charpak studied the operation of a pulsed spark counter with mesh electrodes (see [Charpak, 1963, 664-669]). In particular, he experimented with the duration of the high-voltage pulse.

His studies revealed that when the pulse duration was less than a few microseconds, sparks developed along the tracks of charged particles. However, with longer pulse durations, spurious sparks appeared. Charpak attributed this to discharge propagation caused by UV emission and electron diffusion (G. Charpak, 1986, private communication*)*.

In the late 1970s, he revisited this effect and began operating the detector at a relatively low constant voltage in proportional mode, which reduced the probability of spurious pulses. This modified detector became known as the parallel-plate avalanche chamber (PPAC).

The connection between this detector and the MWPC will be discussed in the next chapter.

3.3. Concluding remarks

At first glance, the concept of the MWPC may seem simple. However, as is often the case in science, an idea appears obvious only after someone formulates it for the first time. A classic example is Einstein's photoelectric effect equation: it consists of just three terms and follows the basic principle of energy conservation. Today, it seems straightforward, yet no one had articulated it before Einstein.

The same holds true for Charpak's MWPC and mesh PPAC. Once introduced, their concepts and designs became almost self-evident. What ultimately matters is the impact of an idea on science and technology. As this book will demonstrate, Charpak's invention had a profound and far-reaching influence.

A key factor in the tremendous success of MWPCs was the development of their readout system using induced signals on segmented electrodes. Today, this method is widely used not only in gaseous detectors but also in solid-state and vacuum-based detectors.

CHAPTER 4

MWPC CONDITIONING

Abstract

As with any device, the initial design of the MWPC required significant refinement to achieve optimal performance as a detector. For educational purposes, it is essential to outline the primary operational challenges encountered with this detector and the solutions implemented to address them

4.1. Initial problems with MWPC

The initial operation of the first MWPC designs was challenging. They frequently exhibited spurious sparks and noise pulses. To address this, a special procedure known as "training" was employed, during which the MWPC was subjected to gradually increasing high voltage over several days. In some cases, the wires were also heated by a high electrical current passing through them. This process aimed to burn off dust and microparticles inside the detector that were causing the discharges

Even after this treatment, the maximum achievable gas gain, A_{max}, in the first MWPC designs remained limited to 10^3-10^4-, which was several orders of magnitude lower than that of single-wire counters (see the previous paragraph). When $A>A_{max}$, sparks or continuous dark currents would appear. However, this gain was still sufficient to detect minimum ionizing particles (MIPs), which defined the initial applications of MWPCs. They were successfully used for imaging particle tracks [Charpak, 1995, 1-650].

Increasing A_{max} was crucial, as it opened new avenues for applications. For example, Seguinot and Ypsilantis proposed utilizing the MWPC's imaging capabilities to visualize Cherenkov rings [Seguinot, 1977, 377-391]. This technique provided a powerful method for particle identification within a specific momentum range.

To achieve this, they proposed adding vapours with low ionization potential, such as benzene, to the standard gas mixture used in MWPCs. The ionization potential of benzene is $E_b = 9.15$ eV. When the energy of Cherenkov UV photons E_{ch} exceeds this threshold $E_{ch} > E_b$, the vapour undergoes photoionization, generating free photoelectrons inside the MWPC. These electrons drift toward the nearest anode wire, where they initiate avalanches upon nearing it. The avalanche position is determined using one of the methods described in the previous paragraph, enabling the visualization of the Cherenkov ring.

However, at a gas gain of 10^3 and with the electronics available at that time, the efficiency of single photoelectron detection was extremely low. Indeed, when the number of primary electrons $n_0=1$, the avalanche pulse-height distribution S_a, follows a characteristic exponential shape (see, for example [Breyer, 1973 91-93]):

$$S_a = \{\exp(-qa/A)\}/A \quad (4\text{-}1),$$

where q_a is the charge in the avalanche and A is the mean charge or gas gain.

What could be recorded by the electronics? Only pulses exceeding the threshold, which was determined by the noise level (see Fig. 4-1).

Therefore, the detection efficiency of single electrons is given by:

$$\eta_{se} = ({}_T\int^{\infty} S_a dq)/({}_0\int^{\infty} S_a dq) \quad (4\text{-}2)$$

where T is the electronic threshold.

At that time, the only practical way to achieve high efficiency was to increase the gas gain of the detector. But how could this be accomplished given the challenges mentioned earlier?

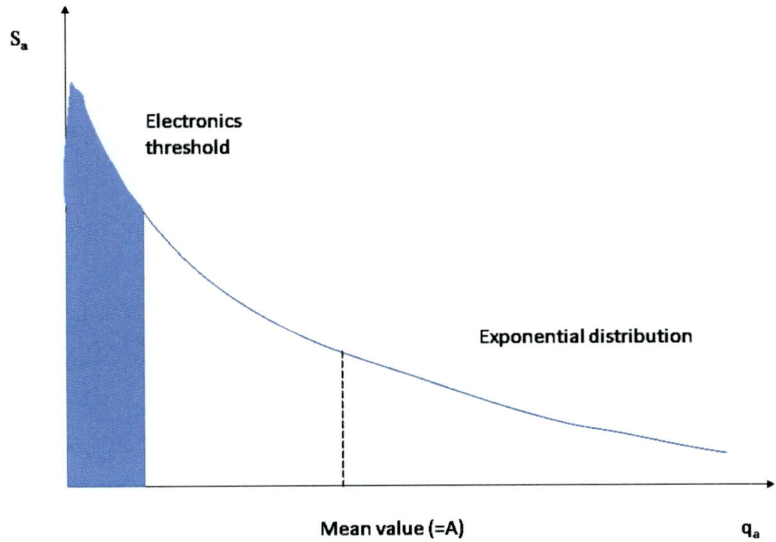

Figure 4-1. Schematic representation of the single-electron pulse-height distribution. Only pulses with amplitudes exceeding the electronic threshold can be detected.

4.2. A multistep detector

Once again, a brilliant solution to this problem came from Charpak and his close collaborators, who introduced the concept of a multistep detector [Charpak, 1978, 523–528]. This idea was inspired by earlier observations reported in the works of [Fischer,1972, 401–404] and [Bohmer,1973, 157–163].

These authors introduced the concept of hybrid detectors: a parallel plate structure with mesh or grid electrodes, combined with a drift space followed by a wire-type spark counter (see Fig. 4-2).

The operation of the chamber is as follows: Primary electrons, generated in the proportional gap by a particle track, drift toward the high-field region at the anode wires, where they undergo avalanche multiplication. These avalanches spread around the wire, with a portion of the electrons penetrating into a secondary region known as the "delay gap." The drift time of these electrons depends on the gap width, the gas composition, and the field strength.

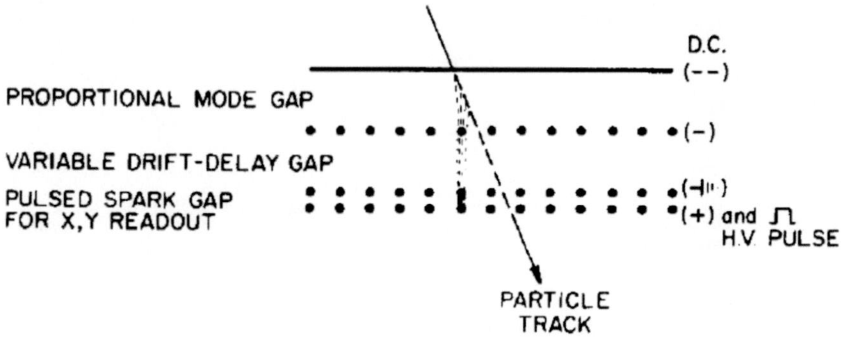

Fig. 4-2. Hybrid Chamber Diagram – The avalanche generated on the second electrode from the top (wire grid) drifts into the final gap, where a gated high-voltage pulse triggers a spark for readout (reproduced from [Charpak, 1978, 1-21]).

As described by Fischer and Shibata [Fischer, 1972, 401-404], the drift time was carefully adjusted to compensate for delays in the trigger counters and logic circuits, which were used to select specific types of nuclear interactions in experiments. The electron cloud then advances into the final narrow gap.

If the counter logic selects a particular event, a short-duration high-voltage pulse is applied to the last gap. At this moment, a spark forms, the position of which can be determined using conventional detection methods. The amplitude and duration of the high-voltage pulse are finely tuned to ensure that sparks form only from avalanche electron clouds that originated on the grid and penetrated the spark gap, while preventing sparking from the small number of primary electrons produced by particle tracks.

Initially, Charpak and his collaborators applied this concept, with several essential modifications (see, for example, [Charpak, 1978, 1-21]), for the detection of single electrons. These electrons were generated in early experiments either by a UV lamp Charpak, 1979, 419-433] or by Cherenkov radiation [Charpak, 1981, 387-396].

The breakthrough idea, however, was to combine a parallel-plate avalanche structure with an MWPC, achieving an exceptionally high gain in proportional mode, exceeding 10^6. This multistep MWPC represented the next generation of gas detectors. Unlike spark chambers, it functioned as a high-rate, continuously operating device, marking a significant advancement in the field.

The schematic layout of the multistep detector is shown in Fig. 4-3. This configuration consists of four distinct regions:

1. A drift space.
2. Two parallel meshes separated by a gap of a few millimeters (referred to as a parallel-plate avalanche chamber, or PPAC).
3. A transfer region.
4. A multiwire proportional chamber.

In this device, n_0 primary electrons, generated in the drift space by ionizing radiation, drift toward the PPAC under the influence of the applied electric field. If the electric field inside the PPAC is significantly higher than in the drift region, a considerable fraction of the primary electrons, $k_d n_0$, penetrates into its volume. The electric field within the PPAC is maintained at a sufficiently high level to induce avalanche multiplication. At this stage, the total number of avalanche electrons created in the PPAC, N_{PPAC}, is given by:

$$N_{PPAC} = k_d n_0 A_{PPAC} = kdn_0 \exp \alpha d, \tag{4-3}$$

where A_{PPAC} is the gas gain, and d is the PPAC gap thickness.

If no electric field is present in the transfer gap, all electrons will be collected on the PPAC anode mesh. However, if a positive electric field is applied between the PPAC and the grid or mesh cathode of the MWPC, some fraction of the avalanche electrons, k_t, will diffuse through the PPAC anode mesh into the transfer region, even if the electric field there is much lower than in the PPAC.

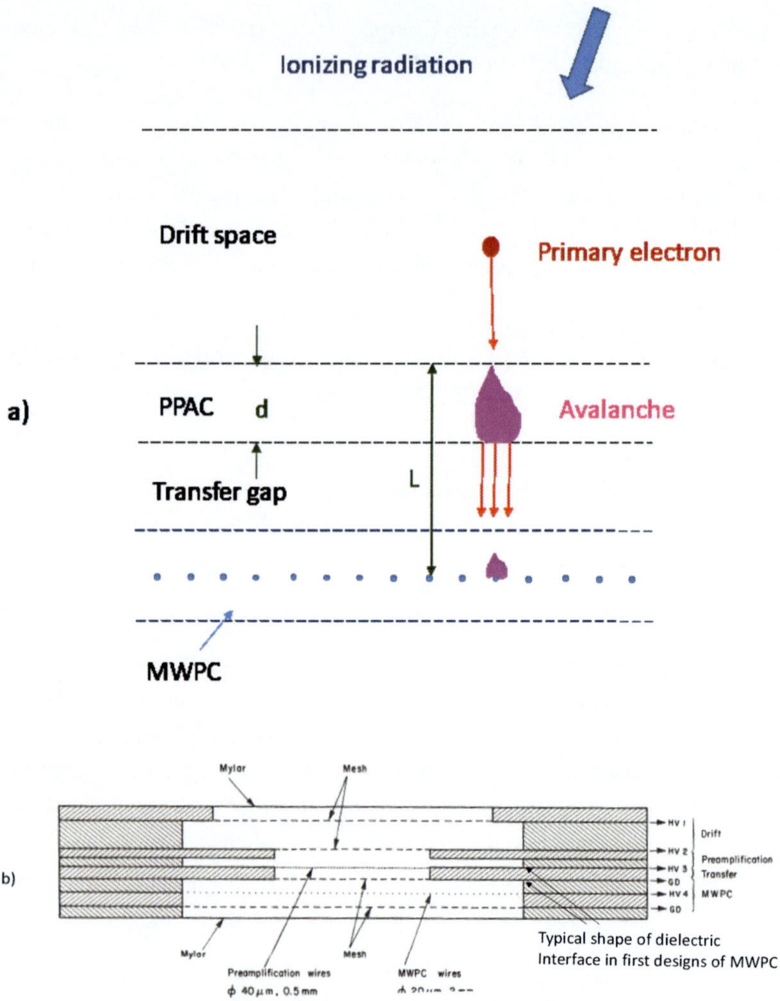

Figure 4-3 a) Schematic diagram of a multistep chamber; b) Technical drawing of a multistep chamber (reproduced from [Charpak, 1978, 1-21], CERN image).

.

After crossing the transfer gap, most of these electrons will reach the MWPC, where they undergo a second stage of multiplication with a gas gain of A_{mwpc} . As a result, the total number of avalanche electrons generated is given by:

$$N_{total} = k_d n_0 A_{PPAC}\, k_t\, A_{mwpc} \quad (4\text{-}4)$$

Thus, the total multiplication factor is:

$$N_{total}/n_0 = k_d A_{PPAC}\, k_t\, A_{mwpc} \quad (4\text{-}5).$$

Typically, in multistep detectors, A_{ppac} is set at approximately 10^2, and A_{MWPC} is close to 10^4. Considering that the coefficient k_d is less than 1, the total gain is around 10^5.

Why is the transfer gap necessary?

Charpak's reasoning behind this was as follows: UV photons from avalanches in the MWPC, due to the optical transparency of the mesh or grid electrodes, may generate photoelectrons outside the fiducial volume. The transfer gap serves to minimize their fraction, preventing them from reaching the drift region.

This relationship can be expressed as:

$$N_{drift} = N_{pm} \exp(-k_{abs} L) \quad (4\text{-}6),$$

where N_{pm} represents the number of UV photons emitted by the MWPC capable of inducing photoionization, and k_{abs} is the effective absorption coefficient of these photons (see Fig. 4-3a).

What about photons from the PPAC generated by avalanches?

These photons can penetrate the drift region and generate photoelectrons, which then undergo full multiplication (see formula 4-5).

Notably, after the initial tests with benzene, researchers identified and utilized several other vapours with even lower ionization potentials as photosensitive elements in multistep chambers. For details, see [Francke, 2016, 1-562]. Examples include:

- Trimethylamine (TMA) – Ionization potential: 7.8 eV
- Triethylamine (TEA) – Ionization potential: 7.5 eV
- Ethylferrocene (EF) – Ionization potential: 6.08 eV
- Tetrakis(dimethylamino)ethylene (TMAE) – Ionization potential: 5.36 eV.

Thus, the question raised earlier becomes critical in validating the multistep detector concept.

Charpak's intuitive idea was that if A_{PPAC} is much smaller than A_{MWPC}, the contribution of these photons remains minimal. It is worth noting that many researchers specializing in the study of creativity in genius scientists suggest that most discoveries and inventions stem from a mysterious sense of intuitive guidance. For further reading, see, for example: [Petravi, 2016, 1-12] and Popva Maria, "The art of scientific investigation: how intuition and the imagination fuel scientific discovery and creativity;" https://www.themarginalian.org/2012/06/01/the-art-of-scientific-investigation-beveridge-2/

Only after the emergence of an intuition-based idea does a scientist begin searching for its logical support or experimental verification.

The same was true for Charpak's hypothesis that photon feedback from avalanches in a PPAC is relatively weak. At the time of the multistep detector's invention, there was no scientific basis for this assumption. However, Charpak later initiated systematic studies of avalanche emission from the PPAC (see [Peskov, 1989, 547-556], [Fonte, 1989, 547-556], [Fonte, 1991, 91-110]).

The key findings can be summarized as follows:

In noble gas mixtures containing photosensitive additives with significantly lower ionization potentials than noble gas atoms (e.g., TMA, TEA, TMAE), electrons primarily excite and ionize these additives when subjected to a relatively low electric field (~1 kV/mm) at one atmosphere. This results in correspondingly low gas gains (~10-100). Low-energy photons emitted during the deactivation of these excited states are unable to ionize the gas mixture. However, some dissociation fragments emit photons with energies exceeding the ionization potential of the original molecules, thereby enabling photoionization of the mixture [Peskov, 1989, 547-556], [Fonte, 1989, 547-556]].

The probability of this process is quite low and decreases rapidly with the concentration of photosensitive additives or the addition of another quenching gas, such as CH_4. The excitation and ionization of noble gas atoms typically occur only at electric fields of $E \geq 10^4$ V/mm, a condition usually met near the anode wires of an MWPC

Thus, as evident, Charpak's intuitive assumptions about the PPAC were later confirmed experimentally. However, this validation came long after multistep detectors had already demonstrated excellent performance in beam test conditions [Mangeot, 1983,79-91].

4.3. Improved designs of MWPC

At the outset, it appeared that the multistep design developed by Charpak and his team was the only viable solution for achieving high gas gains in MWPCs.

A significant breakthrough in this area was later achieved by a Russian research group, which utilized an MWPC for plasma diagnostics in the UV spectral region (see [Bogomolov 1978, 778-783] and [Peskov, 1980, 1-410].

In an effort to achieve the high gains required for detecting single photoelectrons, they investigated various detector designs, focusing on optimizing the dielectric interface between the anode and cathode planes, as well as the materials used in its construction.

Their studies revealed that in the "classical" MWPC configuration (see Fig. 4-4a), which was widely used at the time, spurious pulses began to appear even in the absence of external ionizing radiation at gas gains typically below 10^3. The rate of these pulses and the voltage threshold of their appearance depended on the dielectric interface materials.

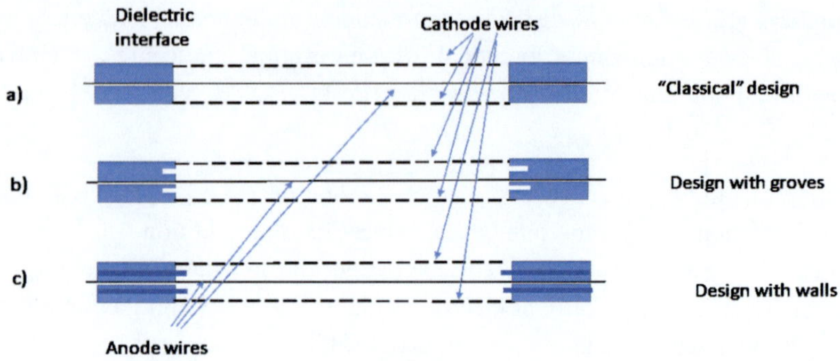

Figure 4-4. Evolution of Dielectric Interface Designs Between MWPC Anode and Cathode Planes: a) "Classical" Structure: This early MWPC design featured G-10 frames to which anode and cathode wires were attached. The dielectric interface between the anode and cathode had a smooth, flat surface. b) Modified Design with Grooves: This version incorporated additional thin sheets, which, when assembled, formed grooves across the dielectric surface, effectively separating the anode and cathode planes (Fig. 4.4b,.c) Multilayer Design with Walls: ***A*** *variation of the multilayer design that allowed for the formation of small "walls" (dielectric insertions) instead of, or in addition to, grooves.*

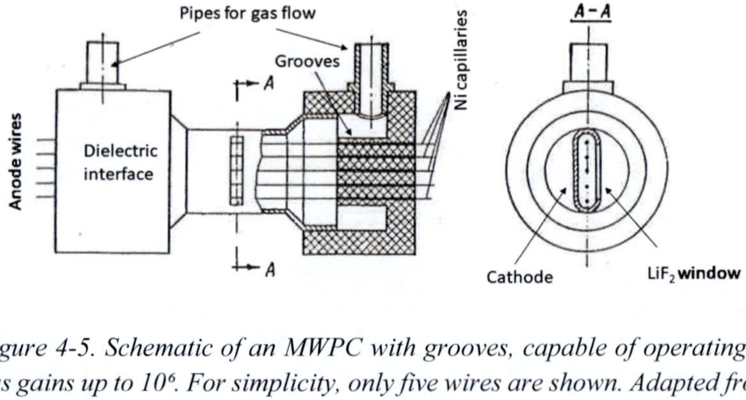

Figure 4-5. Schematic of an MWPC with grooves, capable of operating at gas gains up to 10^6. For simplicity, only five wires are shown. Adapted from [Bogomolov, 1978, 778–783].

As the voltage increased, the rate of these pulses escalated, eventually leading to a transition into discharge

From these observations, the investigators concluded that the pulses and subsequent breakdown were most likely caused by the leakage of electrical charge along the dielectric surface separating the anode and cathode planes. This leakage could manifest as short current pulses.

Based on these findings, an improved MWPC design was developed (see Fig. 4-5 and [Bogomolov, 1978, 778-783]).

This new design introduced two key features:

1. Plexiglass Dielectric Interface: The dielectric interface was made of plexiglass, which significantly reduced the rate of noise pulses compared to G-10, even in the structure shown in Fig. 3-4a.
2. Deep Grooves on the Dielectric Surface: To minimize potential charge leakage along the surface, deep grooves were incorporated into the dielectric interface (see Fig. 4-5).

This design proved to be highly successful: in a mixture of argon and toluene, gains exceeding 10^6 were achieved without the occurrence of any spurious pulses.

The detector was subsequently used in plasma diagnostics to visualize instabilities in the UV spectral region, leading to the discovery of new phenomena through this powerful technique. G. Charpak soon became aware of these developments after receiving a letter from the authors of this work (see details in the section "Some Fun Stories About Charpak"), along with a copy of their paper.

Georges, known for his dynamic nature and keen interest in new advancements, responded almost immediately. This marked the beginning of scientific communication between the two groups. In 1982, he invited one of the authors (V.P.) to present their findings at the Vienna Wire Chamber Conference and to give a seminar at CERN (The European Organization for Nuclear Research). This event introduced the high-energy physics community to the significant new design features of high-gain MWPCs.

In 1984, Charpak visited the P.L. Kapitza Institute for Physical Problems at the USSR Academy of Sciences (see "Some Fun Stories About Charpak"), where he not only observed the detector in operation but also actively participated in some of its tests. This was characteristic of Georges, he enjoyed personally experimenting with new devices.

It is difficult to determine whether the Russian work played a decisive role or if CERN specialists arrived at the same design features independently. However, what ultimately matters is that grooves were soon incorporated into multistep detectors (see Fig. 4-4b) and eventually became a standard feature in nearly all MWPC designs. This innovation enabled gas gains of 10^5-10^6. In some designs, dielectric wall insertions were used instead of, or in addition to, grooves (see Figs 4-4 and 4-6).

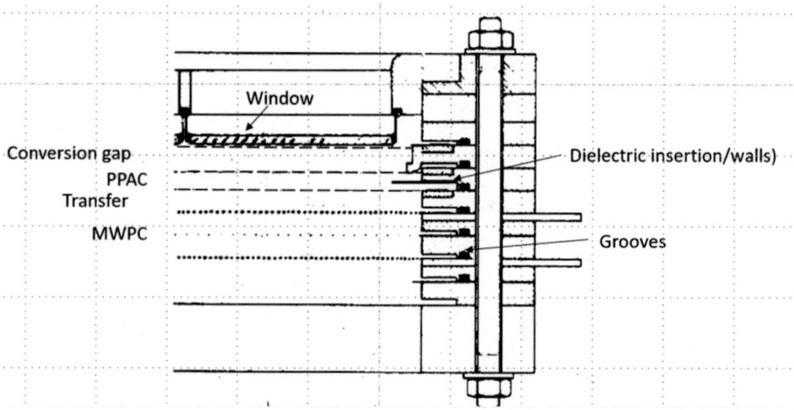

Figure 4-6. An improved design of a multistep detector incorporating grooves between the anode and cathode plates (adapted from [Bouclier, 1983, 403-423)].

The factors limiting the maximum achievable gains, A_{max}, in MWPCs are now well understood. They are outlined below.

In an ideal MWPC, where wires are well-stretched, the anode-cathode dielectric interface is properly designed, and all construction materials are free of contaminants, A_{max} is determined by feedback processes. At

$$A_{max}\gamma = 1 \qquad (4\text{-}7),$$

where γ is the probability of secondary processes (see the first paragraph), a corona discharge occurs. The typical corona current is around ~μA, and, as a rule, this "mild" discharge does not damage either the detector or the front-end electronics.

In contrast, sparks, if they occur, can have currents reaching ~A, making them highly destructive. In real-world conditions, occasional or even continuous sparking can arise if the MWPC is not designed or assembled properly. Sparks may be triggered by loose wires, charge leakage along the dielectric interface, or contamination from dirt, microparticles, or dust introduced through mishandling.

At the beginning of the MWPC era, before clean rooms were routinely used for detector assembly, a special conditioning procedure was employed to address these issues. As mentioned in Section 4.1, high voltage (HV) was applied incrementally, allowing sparking at each stage until all sparks had ceased. This process gradually burned away dust particles, ensuring the MWPC became fully operational.

Nowadays, with extensive experience accumulated, teams involved in designing, manufacturing, and operating MWPCs typically take all necessary precautions. As a result, MWPCs generally function flawlessly in real experiments.

MWPCs and their successors, the drift chamber and time projection chamber, are now regarded as among the most successful and reliable detectors of ionizing radiation.

Concluding remarks

With an improved dielectric interface design between the anode and cathode planes, gas gains of up to 10^5–10^6 became achievable, enabling the detection of single primary electrons with nearly 100% efficiency. This advancement significantly broadened the applications of MWPCs, particularly in Cherenkov photon detection.

Optional Advanced Reading

Further Investigations into Charge Leakage Along Dielectric Surfaces

The physics underlying the origin of the specific spurious pulses described above, which can be suppressed by grooves, remained largely unexamined for many years. Interest in this phenomenon resurfaced much later, around 1990, following the invention of the microstrip gas counter (MSGC) by A.

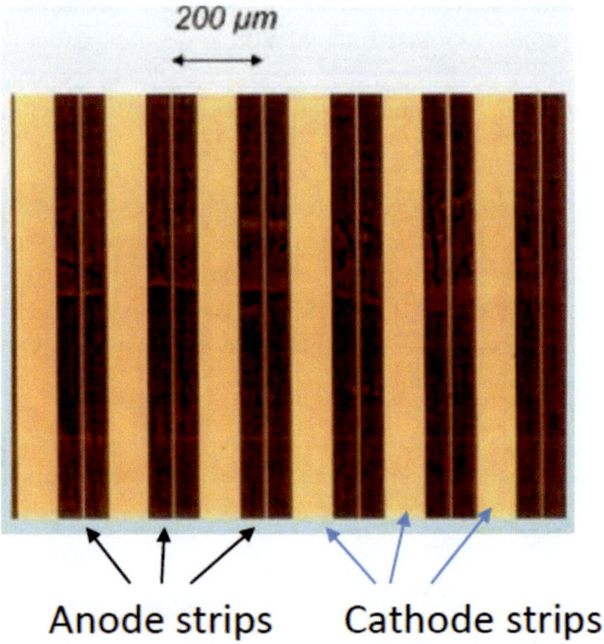

Figure 4-7. Magnified photograph of the MSGC surface, illustrating alternating anode and cathode strips. Image source: https://slideplayer.com/slide/13141835/.

Oed [Oed, 1988, 351-359]. Similar phenomena were observed in this detector.

The description of the MSGC, along with other types of gaseous micropattern detectors, can be found elsewhere (see, for example, [Francke, 2014, 1-303]).

In the MSGC, metallic strips fabricated using microelectronic technology on a dielectric substrate (such as glass or Kapton) replace the anode and cathode wires used in MWPCs (see Fig. 4-7).

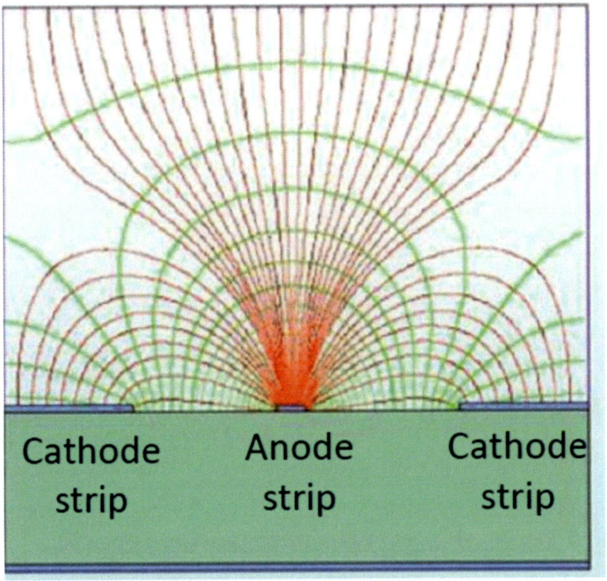

Figure 4-8. Computer-simulated electric field (red) and equipotential lines (green) generated by applying a voltage difference between the anode and cathode strips of the MSGC. Adapted from: https://www.slideserve.com/jeroen/title.

The anode and cathode strips are arranged in an alternating pattern, with gaps typically below 100 micrometers. Essentially, the MSGC can be considered a "small-pitch MWPC" manufactured using advanced fabrication techniques. Notably, the equipotential lines near the anode strips form circular patterns (see Fig. 4-8), indicating that, like MWPCs, the MSGC operates as an array of independent avalanche detectors.

When Charpak first examined the concept of this new device, during a meeting at CERN in 1989 with H. Sipila (one of the authors of this book, V.P., was a witness), he immediately pointed out that the dielectric surface between the anode and cathode strips would be a source of electrical issues, such as charge buildup and breakdowns. Drawing from his experience with MWPCs, Georges knew that without features like grooves or micro-walls, the dielectric surface would be electrically vulnerable, especially when very short, as in MSGCs (see Fig. 4-7).

Subsequent developments confirmed his concerns. For instance, Fig. 4-9 displays spurious pulses observed in MSGCs at voltages near breakdown. Studies conducted in [Peskov, 1997, 89-93] and [Peskov, 1997, 243-260] revealed that these pulses originate from streamers attached to the dielectric surface.

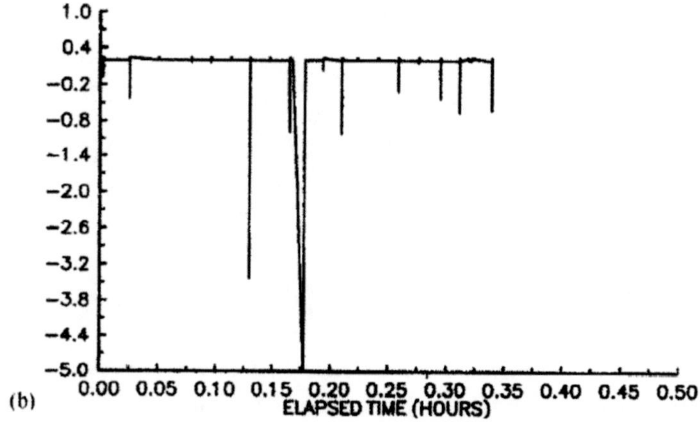

Figure 4-9. Streamer pulses recorded by a computer in the absence of any external radiation source. Reproduced from [Peskov, 1997, 243-260] with permission from Elsevier.

For dielectric substrates, calculations reveal regions of extremely high electric field where the metal electrode structures contact the substrate. These regions are particularly favourable for initiating avalanches near the surface and triggering streamer formation. Field calculations further indicate that when streamers attach to a dielectric surface, they generate a strong local field near their heads, which sustains their propagation (see Fig. 4-10).

These streamers closely resemble the well-known "gliding" discharges that occur across the surface between anode and cathode strips [Uhlig, 1991, 1-220]. Unlike traditional three-dimensional discharge gaps, gliding discharges are two-dimensional. Due to the extreme sharpness of the surface space charge layer at their edges, the transition from surface avalanche to spark-type breakdown occurs at much lower electric fields than in conventional three-dimensional discharges [Uhlig, 1991, 1-220].

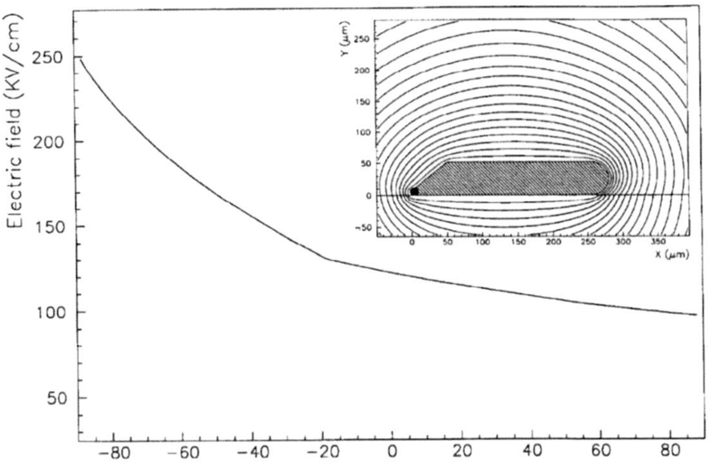

Figure 4-10. Electric field strength around the tip of a streamer. The inset shows a detailed map of equipotential lines for a streamer near the anode of an MSGC in the presence of a surface. Reproduced from [Peskov, 1997, 89-93] with permission from Elsevier.

To highlight the significance of the dielectric interface shape between the anode and cathode strips in MSGC performance, researchers investigated various MSGC-like detector designs in addition to the standard MSGC (see Fig. 4-11 and [Peskov, 1998, 244-248].

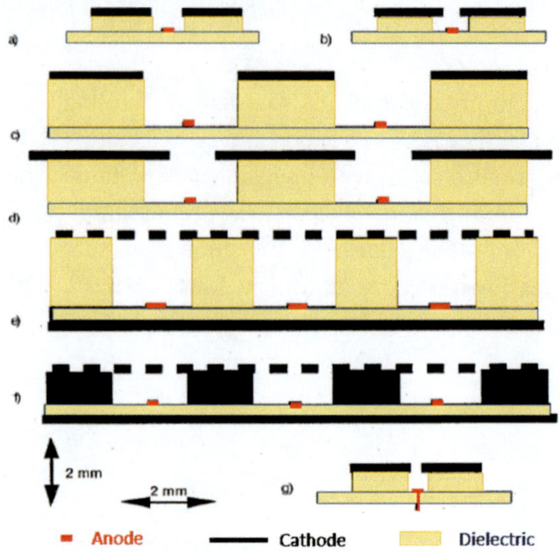

Figure 4-11. Designs of improved microstrip (a-f) and pixel-type (g) detectors. Metallic components are shown with thick lines or dark shading, while the unshaded areas represent the dielectric. For better visualization, the scales of the anode strips and cathode meshes are exaggerated. Adapted from [Peskov, 1998, 244-248].

It has been demonstrated that in standard small-pitch MSGCs (<1mm), the maximum achievable gas gains are limited by the formation of surface streamers, which, at a critical voltage, transition into violent discharges.

Geometrical modifications to the interface, as shown in Fig. 4-11, enabled an increase in the maximum gain by a factor of 5-10.

The results of surface streamer studies in MSGCs appear to be applicable to MWPCs, helping to explain the origin of spurious pulses and the role of grooves in preventing them. In fact, these studies confirm the early qualitative understanding of the impact of charge leakage along the dielectric surface on the performance of MWPCs.

CHAPTER 5

MWPC AND PPAC DESCENDANTS

Abstract

This chapter explores the key descendants of the MWPC: the drift chamber, the time projection chamber, and the photosensitive multiwire chamber. It also highlights Charpak's pivotal contributions to these advancements.

Introduction

As is often the case, any breakthrough in science and technology sparks a wave of subsequent advancements. G. Charpak's invention of the MWPC was no exception, it initiated a series of developments, primarily driven by the high-energy physics community.

Many of these advancements were inspired by Charpak's wealth of innovative ideas, which he continuously generated and generously shared with colleagues, both in private discussions and at various conferences.

Consider just a couple of examples.

In 1969, shortly after inventing the MWPC, Charpak presented the first results obtained with a modified version of his detector at the Dubna International Seminar. This work, conducted in collaboration with T. Bressani et al., focused on track localization using a drift chamber. Their findings were published in [Bressani, 1969, 275-281]. The essence was: "A drift space was added to a proportional chamber. The electrons liberated in the gas of the drift space under the influence of a uniform electric field, pass through a grid into the proportional chamber and are detected there. The time taken to cross the drift space gives measurements of the position".

The schematic drawing of this device is shown in Fig. 5-1, and its photograph is shown in Fig. 5-2.

This brief publication sparked a new direction in detector development, leading to the creation of drift chambers and time projection chambers, as described in Sections 5.1 and 5.2.

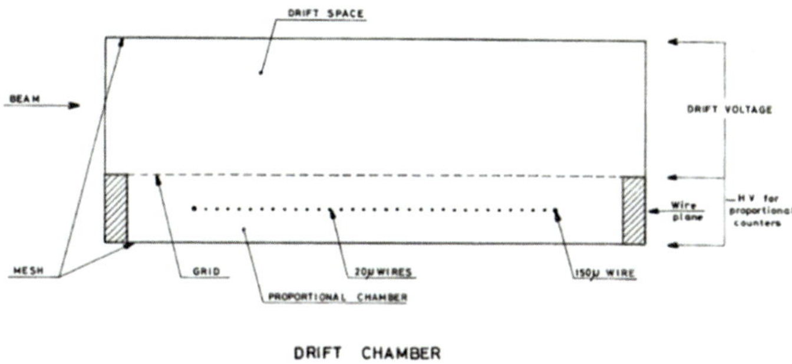

Figure 5-1. First design of a drift chamber, developed and tested by Charpak's team. Reproduced from [Bressani, 1969, 275-281]. CERN image.

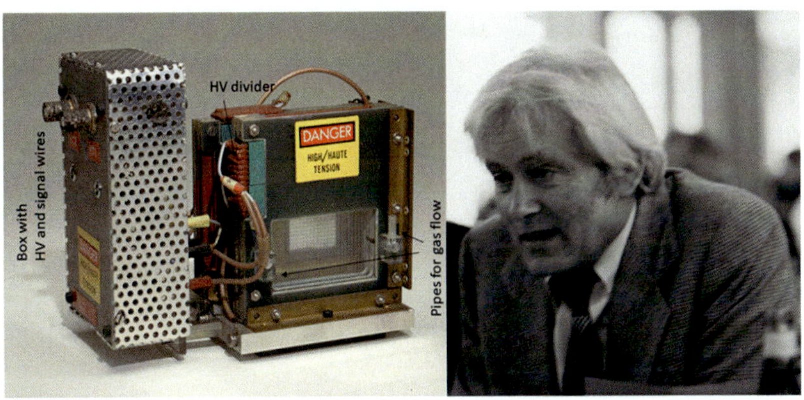

Figure 5-2. Photograph of Georges Charpak (CERN image) and his first prototype of a drift chamber (adapted from https://universemagazine.com/en/george-charpak-a-nobel-laureate-with-ukraine-in-his-heart/.

Another example: As briefly mentioned in the previous paragraph, in 1977, Seguinot and Ypsilantis [Seguinot, 1977, 377-391] introduced a novel technique for particle identification, Ring Imaging Cherenkov (RICH) detector. This method employed a MWPC filled with a gas mixture containing a low-ionization-potential component, such as benzene, which could be ionized by Cherenkov UV photons. Their approach involved visualizing Cherenkov light as cones or rings, depending on the detector geometry, to determine particle velocity (see details in the next chapter). However, at the time, MWPCs operated at relatively low gas gains, insufficient for efficient single-photoelectron detection, making implementation challenging.

Further advancements in this field were made with the active participation of Georges, whose contributions were decisive in shaping this technique. The most significant developments include the following:

To address the issue of low gas gain typical of early MWPC designs, Charpak and Sauli introduced the concept of a multistep gas detector, an approach that enhanced overall gain (see the previous Chapter and [Charpak, 1978, 523-528]).

Shortly after this invention, Georges and his team, in collaboration with Ypsilantis, tested the first RICH prototype, which was based on two parallel-plate chambers with wire electrodes operating in a cascaded mode [Charpak, 1979, 419-433]. This particular device was designed to detect single photoelectrons in a sparking regime, with imaging performed using a standard optical camera.

Later, Charpak and his collaborators successfully detected single photoelectrons produced by Cherenkov radiation in avalanche mode using a multistep detector, with an MWPC serving as the final amplifying element [Mangeot, 1983, 79-91].

Thanks to preamplification, the overall gas gain of this system was sufficiently high to detect single photoelectrons with an efficiency of nearly 90%.

To identify the optimal photosensitive substance based on its spectroscopic and chemical properties, Charpak initiated an ambitious research program within his group to explore and evaluate various photosensitive materials, gaseous, liquid, and solid.

Over time, this effort expanded to include many other research groups worldwide. The fascinating results of these studies will be presented in Section 5.3. Here, we simply highlight one key outcome: large-area photosensitive MWPCs have been, and continue to be, successfully employed in numerous high-energy physics experiments, such as DELPHI (Detector with Lepton, Photon, and Hadron Identification) and ALICE (A Large Ion Collider Experiment) - see Chapter 7 for more details.

This chapter reviews the most significant advancements in wire-based detectors inspired by Charpak's 1968 invention of the MWPC, including drift chambers, time projection chambers, and photosensitive wire detectors.

5.1. Drift chamber

In 1971, Walenta and his group implemented Georges's idea of using the drift time of primary electrons, from their point of creation to the anode wires, to determine their third coordinate. The first two coordinates were derived from the induced charge distribution on the segmented cathode electrodes and the anode wire signals.

To achieve this, they proposed a modified MWPC design in which the traditional anode wire plane was replaced with alternating anode and cathode (or "potential") wires (see Fig. 5-3 and [Walenta, 1971, 373-380]).

The key objective was to improve the field line distribution in the central wire plane, where standard MWPCs exhibit regions of low electric field between anode wires (see Fig. 2-1). In particular, the electric field is nearly zero at the midpoint between two wires, a so-called "critical region."

Although the introduction of potential wires did not eliminate these low-field regions, it leveraged the fact that, in certain gas mixtures, the drift velocity of electrons remains nearly constant over a specific range of electric fields (see Fig. 5-4). With proper corrections, the drift time of primary electrons reaching the anode wires could be converted into their third coordinate.

This modified MWPC design became known as the drift chamber.

The next breakthrough came once again from Charpak and his collaborators, who introduced a high-accuracy drift chamber. In this detector, the cathodes consisted of parallel, equidistant wires arranged at progressively increasing potentials (see Fig. 5-5 and [Charpak, 1973 413-426] and [Breskin, 1975, 189-214]).

This modification enabled the creation of a more uniform electric field between the anode and the potential/field wires along the drift space, offering a significant improvement over the Walenta drift chamber.

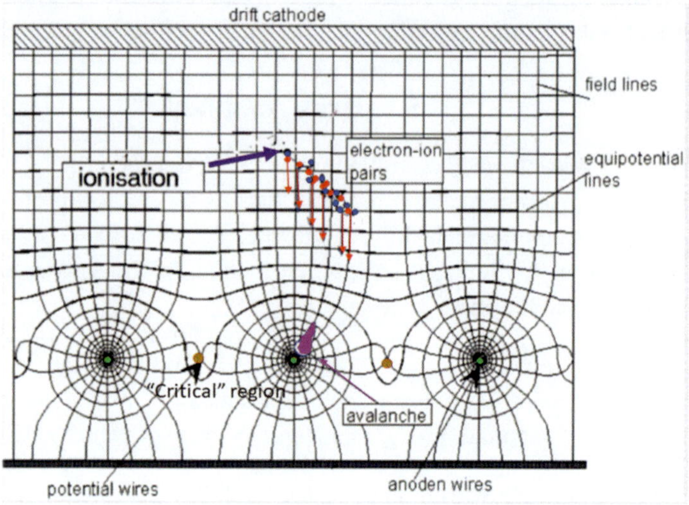

Figure 5-3. Field lines in one of the drift chambers developed by the Helderberg group (note that the distance to the drift cathode is not to scale). As shown, the electric field in the critical region is significantly reinforced compared to the configuration depicted in Fig. 1-14 (adapted from https://indico.cern.ch/event/426015/contributions/1047606/attachments/906077/1278746/DriftChamber_ICFA2005.pdf).

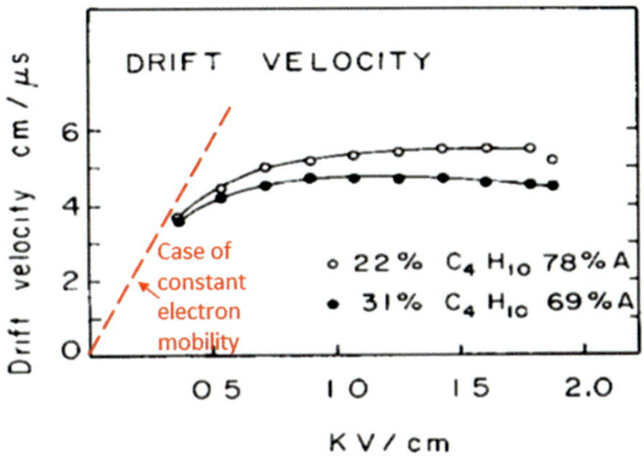

Figure 5-4. Drift velocity as a function of the electric field at atmospheric pressure for two mixtures of Ar and isobutane in different proportions. The drift velocity approaches a constant for fields greater than approximately 500 V/cm (adapted from [Charpak, 1973 413-426]).

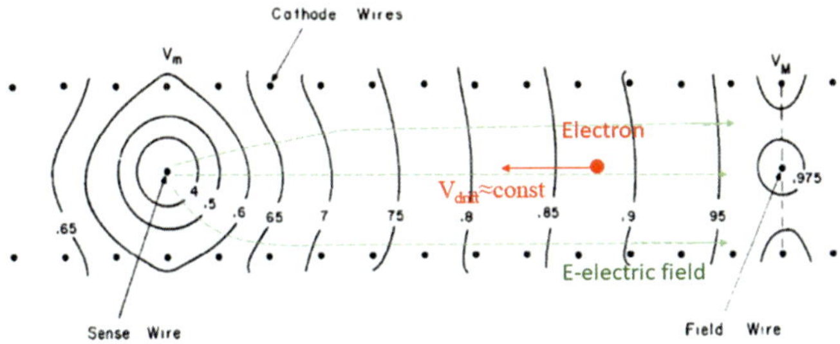

Figure 5-5. Principle of the construction of a high-accuracy drift chamber. The potential on the cathode wires increases uniformly from a minimum ((V_m) near the amplifying wire (located on the left in the figure) to a maximum (V_M) at the end of the drift space. As shown, an almost equipotential field distribution can be achieved in the median plane of the drift cell, even for gaps significantly narrower than the cell size. The addition of a field wire at the centre of the separation between two sense wires helps eliminate the critical low-field region (adapted from [Charpak, 1973, 413-426]).

Later, several other chamber designs, optimized for specific applications, were developed and successfully used in various experiments. One example is a so-called time expansion chamber, a gaseous tracking detector designed to achieve very high spatial resolution over a short drift distance (mm–cm)[Commichau, 1985, 267-78]. Its key concept is to artificially stretch (expand) the drift time of ionization electrons, allowing longitudinal diffusion and the drift-velocity mapping to convert small spatial separations into larger, more easily measurable time differences. Another example is proportional drift tubes, a modular version of a multiwire drift chamber (see [Fitch, 1994, 373-375]), in which the drift field is formed similarly to that of the high-accuracy drift chamber (see Fig. 5-4). A detailed description of these new designs is beyond the scope of this book. Readers seeking comprehensive information on these detectors can refer to [Blum, 2008, 157-180]. Drift chambers typically measure two spatial coordinates per hit directly: one coordinate is determined by which sense wire fired, providing the position perpendicular to the wire plane, and a second coordinate is obtained from the electron drift time to the wire, which gives the distance from the wire.The third coordinate (along the wire) is not intrinsic to a simple drift chamber. It is usually obtained by using multiple layers with wires at different angles (stereo layers) or by combining hits from several chambers through track fitting. Nowadays, most high-energy physics experiments utilize a different type of drift chamber, which is described in the next paragraph.

5.2. Time projection chamber

The further development of Charpak's idea, utilizing electron drift time to determine their 3D position, was pioneered by Nugren, who introduced what later became known as the Time Projection Chamber (TPC; recall the earlier abbreviation) [https://s3.cern.ch/inspire-prod-files c/c86794b79f7db 5d181368c711c0a13f2]. This concept is illustrated in Fig. 5-6, a hand-drawn image taken from his proposal. In the proposed version, the TPC features a cylindrical drift space and a 2D position-sensitive end-cap detector. Its operating principle follows the concept formulated by Charpak's team in their paper [Bressani, 1969, 275–281]. The key innovation was the introduction of a coaxial magnetic field. While initially

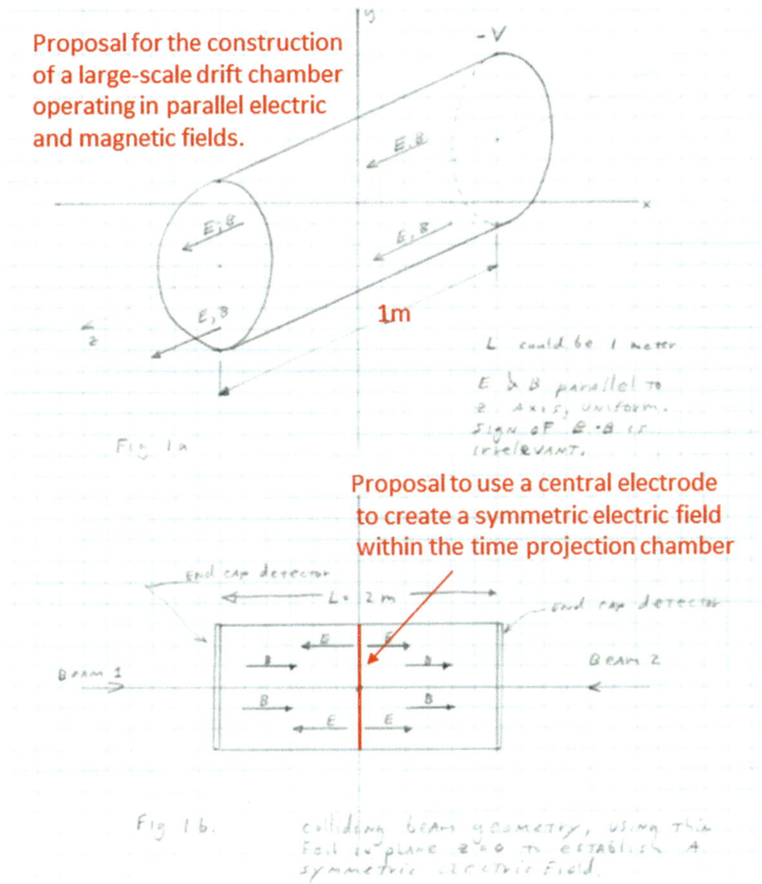

Figure 5-6. Hand-drawn sketch of a drift chamber by Nugren (adapted from [https://s3.cern.ch/inspire-prod-files-c/c86794b79f7db5d181368c711c 0a13f2]). Reproduced with the permission of Dave Nygren.

intended to reduce the diffusion of primary electrons, an even more significant effect emerged: charged particles traversing the gas leave behind an extended trace of ionized gas, preserving a record of their passage. The degree of bending in this trace depends on the particle's charge and momentum, while the ionization density varies based on the particle's identity and energy. The ionization trace then drifts at a constant velocity toward one of the two end plates. In subsequent publications [Marx, 1978, 46–53], the authors proposed using a MWPC as the end-cap detector.

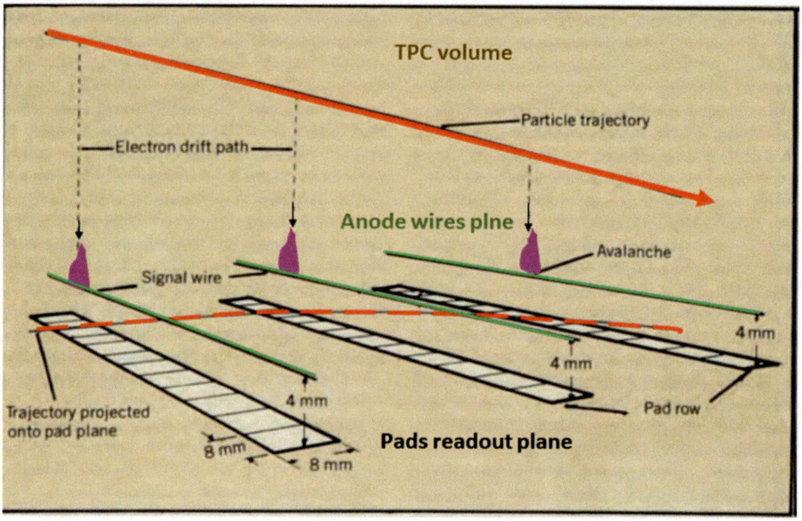

Figure 5-7. Simplified isometric representation of the TPC readout plane, showing 8 mm × 8 mm cathode segments positioned beneath proportional sense wires. The centre of gravity of the signals induced on the cathode segments is used to determine the position of the proportional avalanche along the wire. Adapted from [Marx, 1978, 46–53].

The cathode facing the drift region consisted of a wire grid, transparent to drifting electrons, similar to the design of Charpak's original drift chamber. On the outer side of the MWPC, the cathode plane was constructed from a printed circuit board with metallized pad rows, electrically isolated from one another and oriented parallel to the anode wires, as shown schematically in Fig. 5-7. Each pad was connected to its own charge-sensitive amplifier, with readout managed by a multiplexer. Primary electrons reaching the MWPC were amplified on the anode wires, while the resulting avalanche ions moving toward the cathode induced a positive charge profile.

Using the distribution of induced signals on the pads over successive time slices and the corresponding known drift times, computer programs can generate a 3D image of the track, as well as determine the differential ionization energy loss dE_I/dx. The pad readout, with its large number of independent readout elements, is particularly well-suited for recording high-multiplicity events. All these features make the TPC a unique detector for particle tracking.

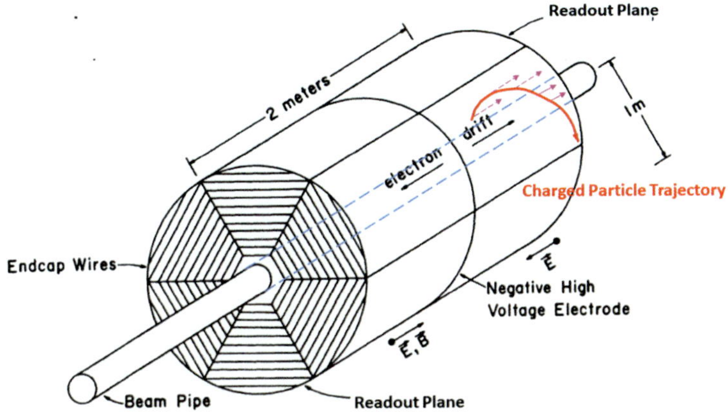

Figure 5-8. Schematic of the first TPC built for the Positron Electron Project-4 (PEP-4) experiment at the Stanford Linear Accelerator Centre (adapted from [Aihara, 1983, 162-168]).

The first TPC, part of the Positron Electron Project-4 (PEP-4) experiment at the Stanford Linear Accelerator Centre, began collecting data in 1983. Its schematic design is shown in Fig. 5-8. The 2-meter-long drift region had a cylindrical shape with a central negative high-voltage plane. The particle beam pipe was positioned along the cylinder's axis. The TPC was housed inside a solenoid, centreed on the interaction point. A negative high voltage applied to the central plane caused electrons to drift in coaxial electric and magnetic fields toward the two readout planes, one at each end, with a drift length of 1 meter. Each endcap featured a MWPC arranged in six sectors, each containing 'azimuthal' anode wires.

One example is the world's largest gaseous TPC built for the ALICE experiment, which is dedicated to heavy-ion physics at the LHC at CERN. More details about this experiment can be found in Chapter 7.

Soon after, additional TPCs were built and successfully used not only for searches for rare and exotic processes with low multiplicities but also in collider and fixed-target experiments with increasing particle fluxes. Both the dimensions and the number of readout channels expanded, enabling the reconstruction of events with extremely high particle multiplicities. This advancement fully leveraged the TPC's exceptional pattern recognition capabilities.

Nowadays the TPC concept is spreading to other fields: liquid and solid-state detectors Probably, the most impressive noble liquid TPC today is a so-called ProtoDUNE –a huge liquid Argon detector. DUNE, the Deep Underground Neutrino Experiment, is dedicated to studying an accelerator-produced neutrino beam generated at Fermilab (Fermi National Accelerator Laboratory). The experiment will include two detectors positioned along the beam's path: one near the source and the other at the Sanford Underground Research Facility, 1,300 km away. The second and larger detector will consist of four $60 \times 12 \times 12$ m^3 TPCs, each containing 10 kT of liquid argon. Research on these baseline detectors is currently underway at CERN. Fig. 5-9 presents images of tracks recorded by this TPC.

A solid-state TPC was developed by the Medipix3 collaboration at CERN [Acharya, 2020, 1-8]. It operates on the same fundamental principles as gaseous and liquid TPCs, with some distinct features. Ionizing particles interacting within the fully depleted zone generate free charge carriers (electrons and holes), which drift under an applied electric field between the common backside and the pixel contacts (a 256×256 pixel array with a 55 μm pixel pitch). As these charge carriers drift, they induce currents at the pixel electrodes, which are subsequently processed, shaped, amplified, and digitized, to determine their 2D coordinates and time of arrival.

By leveraging the known relationship between drift velocity and electric field, the interaction depth can be inferred either analytically (for silicon) or numerically (for CdTe). Studies have demonstrated that a third-coordinate resolution of approximately 30 μm can be achieved in 500 μm thick silicon sensors, while ~60 μm resolution is attainable in 2 mm thick CdTe

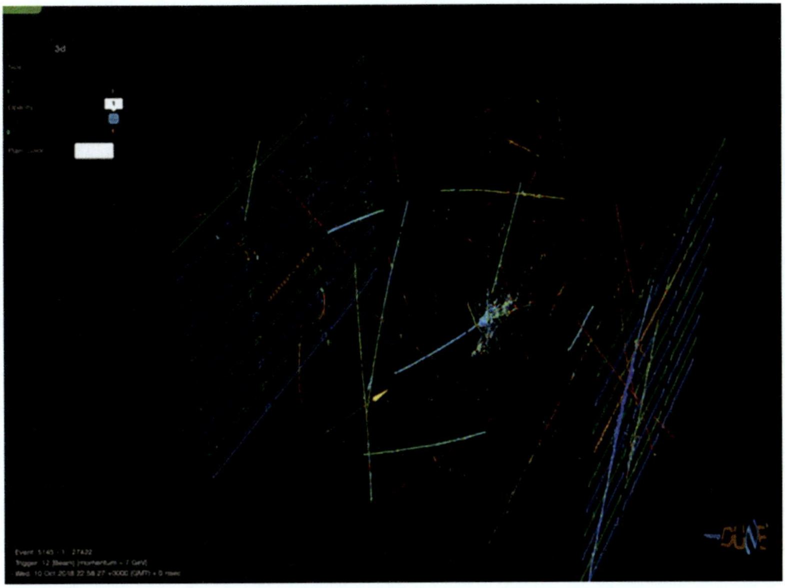

Figure 5-9. A reconstructed test beam π^+ (entering the ProtoDUNE-SP TPC at the yellow arrow) interacting inside ProtoDUNE-SP, accompanied by several coincident cosmic ray muons. Image credit: CERN.

Today, TPCs are still in a developmental stage, with numerous projects either utilizing or planning to explore this technology. New signal readout elements and methods are also being considered, such as micropattern gaseous detectors (see Chapter 10) and optical readout, where particle tracks in the TPC are recorded using fast CCD (charge-coupled device) cameras. Notably, the first attempt to use optical readout for detecting particle tracks in a TPC-like structure was made by the Charpak group [Suzuki, 1988, 237-242].

Given these ongoing advancements, TPCs are expected to remain one of the primary tracking devices and will continue to have a promising future.

5.3. Photosensitive MWPC

Photosensitive MWPCs were independently introduced almost simultaneously by two research groups: one from CERN [Seguinot, 1977, 377-391] and another from the P.L. Kapitza Institute for Physical Problems in Moscow [Bogomolov, 1978, 778-783]. The difference in the submission dates of their articles was only three months.

However, the two groups aimed at different applications. Seguinot's team focused on Cherenkov ring imaging, while the Russian group employed their photosensitive MWPC to visualize plasma instabilities in the ultraviolet (UV) region of the spectrum (see Chapter "Other Applications of MWPCs").

What is a photosensitive MWPC?

A photosensitive MWPC is a detector equipped with a window transparent to ultraviolet (UV) light. It operates in a standard gas mixture, typically argon combined with a quenched gas at a total pressure of 1 atmosphere. To make the detector sensitive to UV light, vapours with a low ionization potential (E_{ips}) are added to the mixture.

When photons with energy $hv>E_{ips}$ enter the chamber, they can ionize these vapours, generating primary photoelectrons. These photoelectrons initiate electron avalanches within the MWPC. The position of these avalanches is then determined using standard electronic methods, as described in Chapter II. In this way, the detector enables electronic imaging of UV light (see [Francke, 2016,1-562] for more details).

As a photosensitive element, Séguinot et al [Seguinot, 1977, 377-391 used benzene vapours (ionization potential E_i=9.15)), while a Russian group employed toluene vapours (E_i=8.8 eV).

The discovery that MWPCs can detect single UV photons sparked a wave of research worldwide. The goal was to identify vapours with ionization potentials lower than that of toluene, thereby extending the spectral sensitivity range, and to test their compatibility with the conventional materials used in MWPCs.

Note that the accuracy of Cherenkov cone/ring reconstruction is directly related to the number of photoelectrons generated in the MWPC per event. This number, in turn, depends, among other factors, on the width of the spectral sensitivity interval Δλ and the shape of the quantum efficiency curve Q(λ) of the photosensitive gases used.

Between 1978 and 1980, several groups successfully employed TMA (trimethylamine, Ei= 7.8 eV) and TEA (triethylamine, Ei=7.5 eV) vapours in photosensitive MWPCs (see, for example, [Peskov, 1980, 507-512], [Charpak, 1978, 523-528], [Charpak, 1979, 419-433], [Alegria Feio, 1980, 473-476], [Seguinot, 1980, 283-298].

The first prototype of a RICH detector using TEA vapours for Cherenkov ring detection was likely described in [Mangeot, 1983,79-91].

Charpak brought together key scientists who had previously worked on photosensitive detectors, including J. Séguinot, T. Ypsilantis, D. Anderson, and others, one of whom was the co-author of this book, V.P. This small but highly effective team made significant contributions in a short period of time.

Anderson brought to CERN a sample of the photosensitive liquid tetrakis-(dimethylamino)-ethylene (commonly known as TMAE, $C_{10}H_{24}N_4$) which he had previously tested while working at Los Alamos. With its low ionization potential of just 5.36 eV, TMAE vapours were highly attractive for use in photosensitive MWPCs. However, they proved to be chemically reactive with oxygen and certain construction materials. As a result, the teams led by Charpak and Séguinot had to overcome various technical and engineering challenges.

Despite these initial challenges, TMAE-filled MWPCs were later successfully employed in RICH detectors across several high-energy physics experiments. The most notable among them was the DELPHI RICH detector, a sophisticated and highly successful apparatus. It remains the largest detector ever built using photosensitive vapours (see the next paragraph for more details).

Another of Georges' collaborators, V.P., brought a bottle of ethylferrocene to CERN. The vapours of this liquid have a slightly higher ionization potential (6.08 eV) and a lower saturated vapour pressure at room temperature compared to TMAE, making them less attractive for RICH applications. However, unlike TMAE, ethylferrocene is non-reactive with air and common materials, which significantly simplifies its handling.

Ethylferrocene was employed in various laboratory projects, including the readout of UV scintillators and prototypes of UV-sensitive, solar-blind flame detectors. A more detailed historical overview of these developments can be found in the introduction of [Charpak, 1995, 1-650].

Following the initial success of photosensitive MWPCs, research groups from around the world, including those from Israel, France, and Britain, joined the effort. As a result of this international collaboration, other vapours were soon tested [Francke, 2016,1-562]

However, the culmination of developments in photosensitive gaseous detectors was the exploration and testing of solid photocathodes in combination with gaseous detectors. This line of research was also initiated by Charpak. The first results in this direction were presented at the Fermilab Conference on Particle Identification (see [Charpak, 1989, 295-308].

Leading specialists in photosensitive detectors attended this meeting, including D. Anderson, A. Breskin, J. Seguinot, and others. As a result, information about this new approach quickly spread within the high-energy physics community, initiating a new cycle of internationally coordinated R&D. Systematic studies of this promising photocathode were conducted at Seguinot's laboratory at CERN [Seguinot, 1990,133-147], while Charpak's group primarily focused on measurements using a PPAC (recall that he had previously used it as a preamplifier for an MWPC) combined with a CsI photocathode. The expectation was to achieve superior time resolution compared to the MWPC, which was subsequently demonstrated experimentally [Charpak, 1991, 63-68], [Anderson, 1992, 626-634]. The Breskin group, building upon these developments, focused on studying MWPCs combined with a CsI photocathode. These three studies [Charpak, 1989, 295-308], [Seguinot, 1990,133-147] and [Dangerdorf, 1990, 322-324] marked the beginning of a new direction in the development of photosensitive gaseous detectors.

A significant achievement was the development of a technology for producing CsI photocathodes with high quantum efficiency and good reproducibility [Peskov, 1995, 48-51]. RICH detectors based on CsI-MWPCs have been successfully employed in several experiments. Currently, the largest of these, with a sensitive area of approximately 7 m^2, has been operating for nearly 14 years in the ALICE detector at CERN's LHC. This long-term operation stands as the strongest evidence to date of the maturity and reliability of CsI photocathode technology. Comprehensive information about gaseous photosensitive detectors and the history of their development can be found in: Francke, 2016, 1-562].

5.4. Optical imaging chambers

In the early years of his scientific career at CERN, Charpak gained experience with spark and streamer chambers, which familiarized him with the optical readout of these devices. This background allowed him to quickly appreciate a paper in which an image intensifier sensitive to visible light was used to record position avalanches in MWPC [Bogomolov, 1978, 639-643].

Charpak made two significant advancements:

1. After initial tests with MWPCs, he shifted his focus to Parallel Plate Avalanche Counters (PPACs), which provided improved two-dimensional spatial resolution. Notably, the original mesh PPACs were not position-sensitive; in this sense, the optical chamber can be viewed as their successor
2. He initiated research and development within his group to identify gas mixtures that produced the highest light output per avalanche charge in the visible or near-ultraviolet spectral regions.

Two groups of such mixtures were identified: any noble gases combined with TMAE (tetrakis-dimethylamino-ethylene) or TEA (trimethylamine) vapours. In some instances, an additional quencher, such as CH4, was added to achieve high gain.

The emission spectra of these mixtures, measured in avalanche mode, are presented in works by [Charpak, 1988, 483-486], [Peskov, 1989, 547-556] as shown in Fig. 5-10.

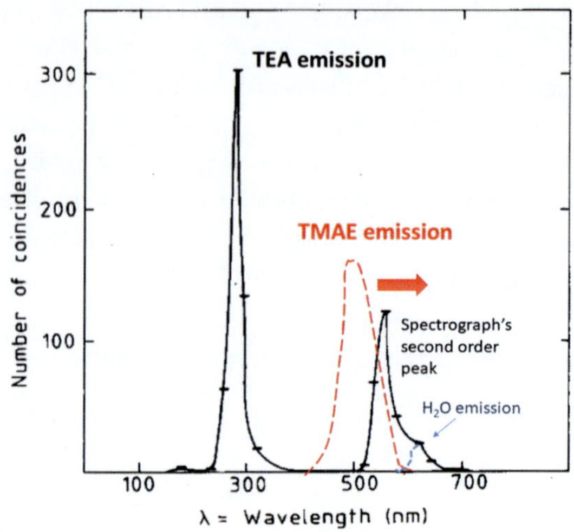

Figure 5-10. Red curve: *Spectrum of visible light emission from avalanches in a gas mixture containing TMAE, with CH_4 added to achieve higher gas gains (adapted from [Charpak, 1988, 483–486]).* Black curve: *Near-ultraviolet avalanche emission spectrum in Ar+TEA mixtures. The peak at 280 nm corresponds to TEA emission, while the peak at 560 nm results from second-order diffraction in the monochromator, providing twice the spectral resolution. A faint trace of water emission is also visible in this region (adapted from [Peskov, 1989, 547–556]).*

Although the TMAE molecules emit in visible spectral intervals, making them easier to record optically, they are chemically aggressive. For this reason, TEA vapours, which are less reactive, are used in most applications. The intensity of the avalanche emission depends not only on the gas composition but also on the gas gain (Fig. 5-11). Note that the emission spectra of these mixtures in the MWPC displayed different characteristics due to the significantly higher electric field.

The schematic of the optical chamber is shown in Fig. 5-12. It contains a PPAC with an extended drift region. Ionizing radiation generates primary electrons that drift toward the PPAC, producing avalanches. The light emitted by these avalanches is detected using an image intensifier connected to either a standard television or a solid-state (CCD) camera. The detector

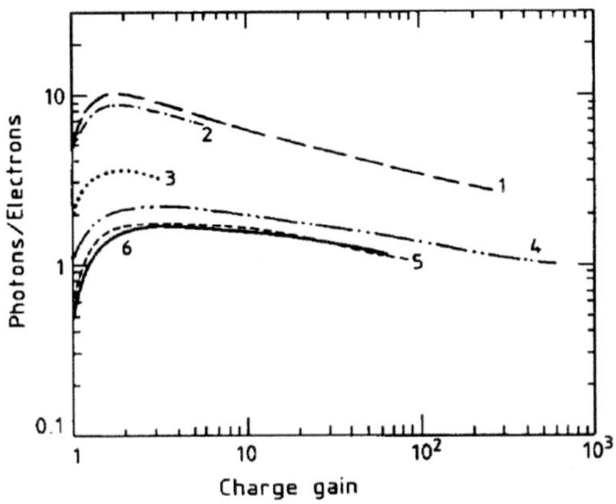

Figure 5-11. Typical dependence of the light-to-charge ratio on the gas gain for different mixtures, all at a total pressure of 1 atm but at various partial pressures of vapours (in Torr):1-Ar+TEA (60),2-Ar+TMAE (0.4),3-Ar+Ethilferrocebe(<10^{-2}), 4-Ar+TEA (0.5),5-"pure" Ar, 6-Ar+$H_2$0 vapours (10). Reproduced from [Charpak, 1988, 483-486] with permission form Elsevier.

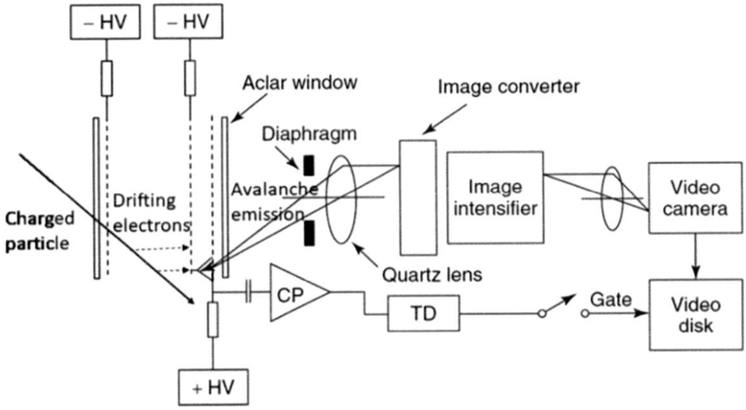

Figure 5-12. Schematics and the principle of operation of the optical imaging chamber. HV stands for high voltage, CP–charge–sensitive amplifier, TD-timing discriminator (adapted from [Charpak, 1988, 483–486]).

can operate continuously or be gated for selective measurements, such as in high-radiation environments. An example of a digitized image recorded with the imaging chamber is shown in Fig. 5-13. The shades of the pixels, originally in color, represent the relative intensity of the recorded light and closely reproduce the typical Bragg energy loss distribution.

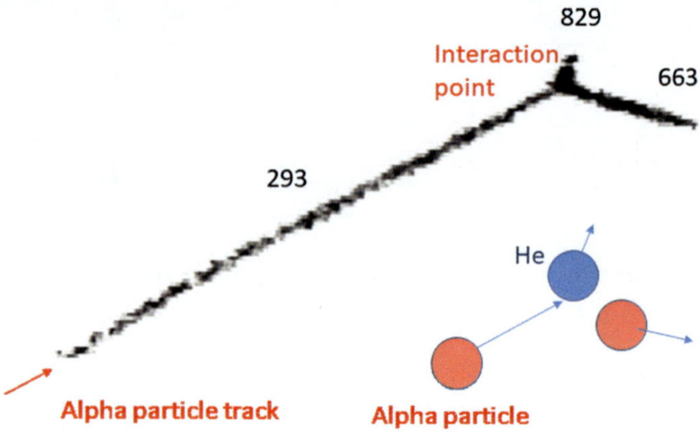

Figure 5-13. Image of an alpha particle interacting within the drift volume of an imaging chamber filled with a He + TEA gas mixture, accompanied by a schematic explanation (drawing below). The track length is approximately 10 cm. The average ionization along the track, indicated by the number in the image, was measured using a CCD (adapted from [Charpak, 1988, 483–486]).

The applications of optical imaging chambers in the past and present are described in Chapter 7.

Concluding remarks

As can be seen from the materials presented in this chapter, Georges not only invented the MWPC and made a decisive contribution to their evolution but also played a major role in the development of their descendants, in particular in drift chambers and photosensitive wire detectors. These new imaging devices dramatically changed the field of detectors of ionizing radiation.

Nowadays drift chambers, TPC, and photosensitive MWPCs are widely used in many high-energy physics. A short review of old and present experiments is given in Chapter 7

Additional reading

For more information on drift chambers and Time Projection Chambers (TPCs), see [Blum, 2008, 157–180]. A detailed discussion on photosensitive gaseous detectors, including their operating principles, designs, and various applications, can be found in [Francke, 2016, 1–562].

CHAPTER 6

THE MWPC FOR PARTICLE IDENTIFICATION, CALORIMETERS AND MUON STATIONS

Abstract

This chapter explores the role of MWPCs in high-energy physics experiments, with a focus on their applications in particle identification, calorimetry, and muon detection systems. It examines their integration within detector architectures and how they complement other subdetectors to enhance tracking precision and event reconstruction. The specific challenges and advantages associated with the deployment of MWPCs in modern detector systems are also highlighted.

Introduction

In the previous chapter, MWPC descendants were described: drift chambers, TPC, and photosensitive wire detectors, all of which made breakthroughs in experimental techniques. Besides this, almost immediately, MWPC began to be successfully used in other components of complex high-energy physics detectors: particle identification systems, calorimeters, and muon stations. Note that, as a rule, such detectors have huge dimensions (tens of squared meters) to trace the evolution in time and space of primary and secondary elementary particles produced in collisions at accelerators. There is a wide variety of such detectors. At collider experiments, most of the modern multilayer detectors have a coaxial cylindrical geometry along the beam axis. Schematically, the cross section of one fourth of their structure is shown in Fig. 6-1. The first coaxial layer is usually occupied by a so-called solid-state vertex detector, which determines the coordinates and angles of particle tracks closest to the interaction point.

Moving far from the vertex, a tracking detector is located next. Usually, for this purpose, drift chambers and TPCs were widely used, although lately, in some modern detectors, such as CMS, silicon trackers are replacing TPCs. After the tracker usually there is a dedicated particle identification system.

There are several methods for the particle identification based on measuring the momentum in a tracking chamber, combined with a measurement of velocity to determine the mass and charge of the particle and, therefore, its identity. For example, as mentioned in the previous paragraph, TPCs, besides tracking, also provide dE_l/dx information that allows the particles identification within a certain momentum range.

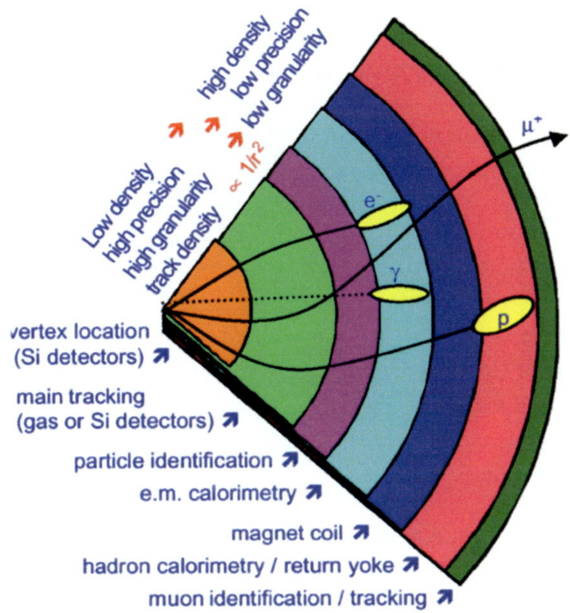

Figure 6-1. Schematic representation of a complex multilayer detectors (Reproduced from https://indico.cern.ch/event/130206/contributions/119291/attachments/91542/130922/NTP_Detectors.pdf, with permission from C. Joram).

To extend this range or select specific particles, various experiments apply additional detection techniques depending on scientific goals; some multilayer detectors may integrate transition radiation detectors (TRD), ring imaging detectors (RICH), and/or time-of-flight chambers. MWPCs are used in TRD and in ring imaging Cherenkov chambers (RICH), which have the property of not being destructive of the particle.

In fact, the subsequent layer is usually an electromagnetic calorimeter that is destructive for electrons or photons. Moving farer from the vertex, hadronic calorimeters are installed that absorb the hadronic component of the event (destructive method). They are followed in the most external layer of the experiment by the muon identification/tracker. Muons interacting weakly with the matter, survive after passing all the previous detector layers.

6.1. MWPC for TRD detectors

The Transition Radiation Detectors (TRD) are based on a physical effect predicted by Ginzburg [Ginzburg, 1945, 353-362] and confirmed later by experiments. Charged particles crossing the interface of two media that have different dielectric constants emit electromagnetic radiation (see Additional Reading 6.8.1).

As an example, we describe here the TRD installed in the ALICE experiment at CERN. It is shown schematically in Fig. 6-2, a drift chamber is combined with a radiator, producing the transition radiation. The radiator is a sandwich of two sheets of rigid foam (polymethacrylimide) filled with a propylene fiber mat.

Two tracks are shown: a pion (blue one), and an electron with momentum in the range of 1–10 GeV/c. The electron produces x-ray transition radiation while traveling through the radiator, whereas the pion does not, as its kinematic factor gamma is below the threshold of ~10^3.

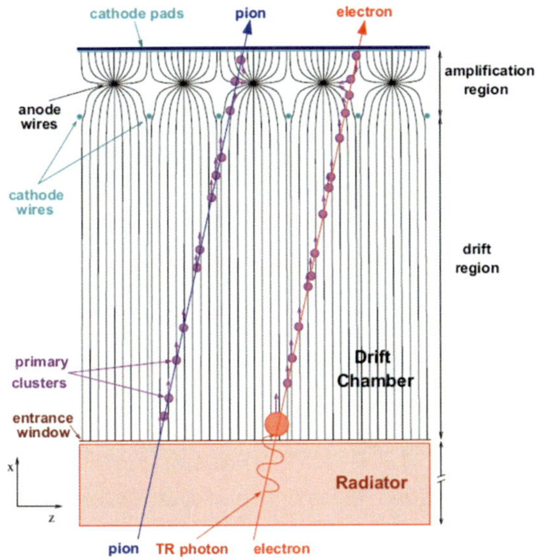

Figure 6-2. Schematic representation ALICE TRD detectors (CERN image).

6.2. MWPCs for RICH detectors

MWPCs have been, and continue to be, used in another particle identification system: the ring imaging Cherenkov detector. The history of the discovery of this radiation and its properties is detailed in Section 6.5.

Here are two main geometries of Ring Imaging CHerenkov detector (RICH) based on MWPCs: with a mirror and with a proximity focusing gap. In a mirror RICH detector, the Cherenkov photons are projected onto the photosensitive MWPC placed at the focal plane (Fig. 6-3). This layout produces a circle with a radius independent of the photon emission point along the particle's track. This geometry is usually used for low refractive index radiators associated to long gaseous radiator, feature needed to produce a sufficient number of photons. In the proximity-focusing design (Fig. 6-4), a thin radiator emits a cone of Cherenkov photons, which traverses a small distance (focusing gat) to expand the Cherenkov pattern that is detected on the photosensitive cathode of a MWPC. For a particle impinging perpendicularly to the radiator, the result is a ring whose radius is defined by the Cherenkov emission angle and the proximity gap. Such a RICH detector still today (2025) is used in the ALICE detector at the LHC.

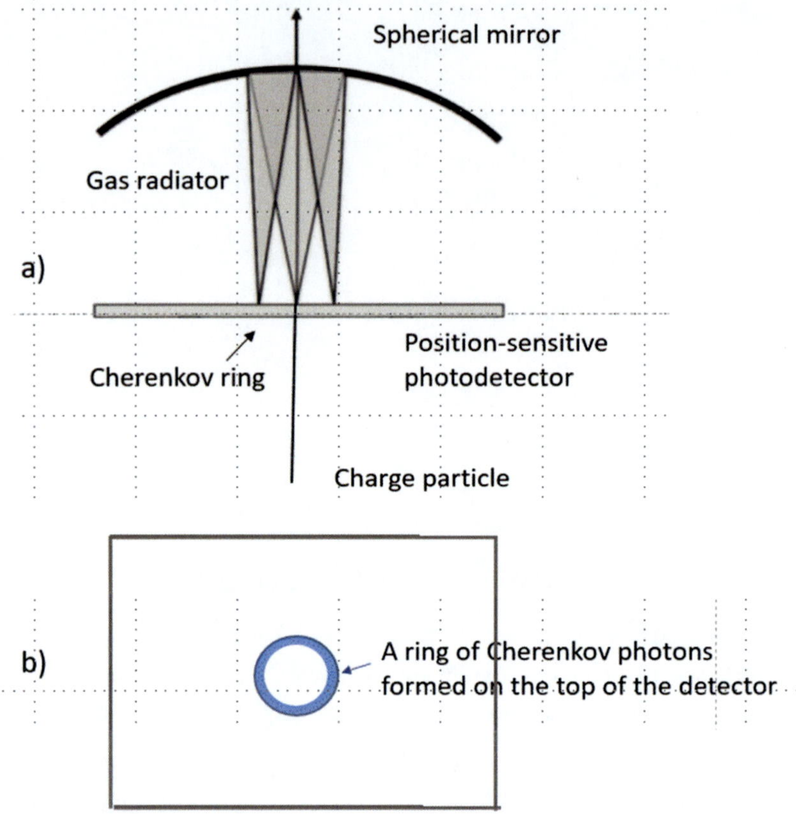

Figure 6-3. Schematic presentation of RICH, consisting of a gas radiator, a focusing mirror, and a position-sensitive detector, which detects a ring of photons (adapted from [Sauli, 2014, 1-497]).

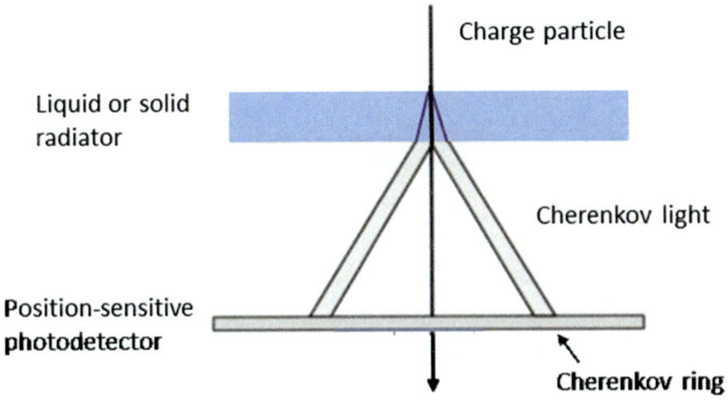

Figure 6-4. Schematic of a proximity-focus RICH detector: Cherenkov light, emitted as particles pass through a thin layer of liquid or solid radiator, forms a conical wavefront in the gas gap and projects a ring of photons onto a photosensitive detector (adapted from [Sauli, 2014, 1-497]).

6.2.1. TMAE based RICH

Relevant examples of the practical implementation of this technique include the SLD RICH at the SLAC Linear Collider and the previously mentioned DELPHI RICH at CERN's LEP (the European Particle Physics Laboratory and the Large Electron-Positron Collider, respectively).

In both cases the MWPCs were filled with TMAE vapour as photosensitive elements to convert the Cherenkov photons. In Fig. 6-5, a schematic drawing of DELPHI RICH is depicted. To extend the particle identification region (particle momentum), it was adopted the combination of two detectors in one assembly: a proximity- focusing RICH with a liquid radiator, and a focusing mirror RICH with a gaseous radiator. Cherenkov photons produced in the liquid radiator are projected onto the drift chamber containing TMAE. To ensure almost their full absorption in the drift chamber volume, the temperature inside the vessel was kept at 40 °C.

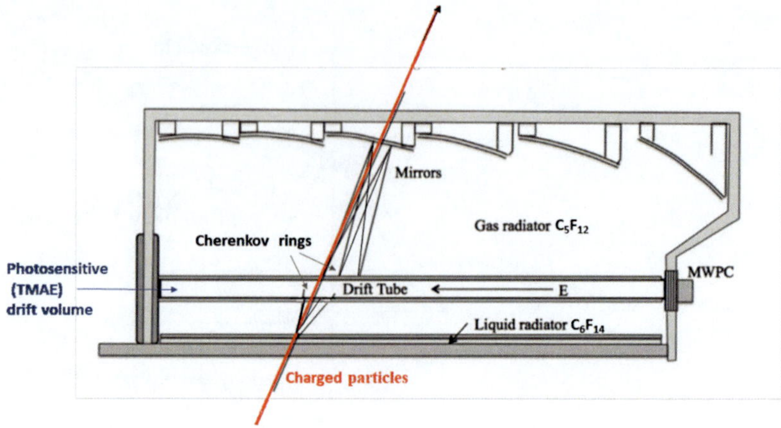

Figure 6-5. Schematic of the DELPHI RICH detector, featuring a liquid radiator (proximity-focus RICH) and a gas radiator with a system of focusing mirrors. Rings of photoelectrons are drifted over distances of up to one meter to the end of the detector, where they are recorded in a multi-wire proportional chamber (MWPC). Adapted from [Arnold, 1988, 289–318].

The drift time of photoelectrons to reach the MWPC (as shown in Fig. 6-6), permits the reconstruction of the Cherenkov pattern.

Because the outside of the barrel-RICH was at room temperature, a system containing heating foils, thermal insulation, and cooling shields was incorporated in the end flanges as well as in the inner and outer cylinders. A major disadvantage of TMAE is its chemical reactivity. It reacts strongly with oxygen and some construction materials, and the oxidation products produced in this reaction have a strong negative influence on the performance of the photon detector. As can be seen, it was a significant challenge to build and exploit such a detector; however, all the problems were ultimately solved successfully.

In 1992, the RD26 collaboration was established at CERN to explore the construction of a proximity-focusing TMAE-based RICH for CERN's upcoming collider - the LHC (recall: Large Hadron Collider). Nonetheless, a more attractive approach was soon selected for the upcoming RICH:

MWPCs equipped with CsI photocathodes. The subsequent paragraph outlines the relevant advantages.

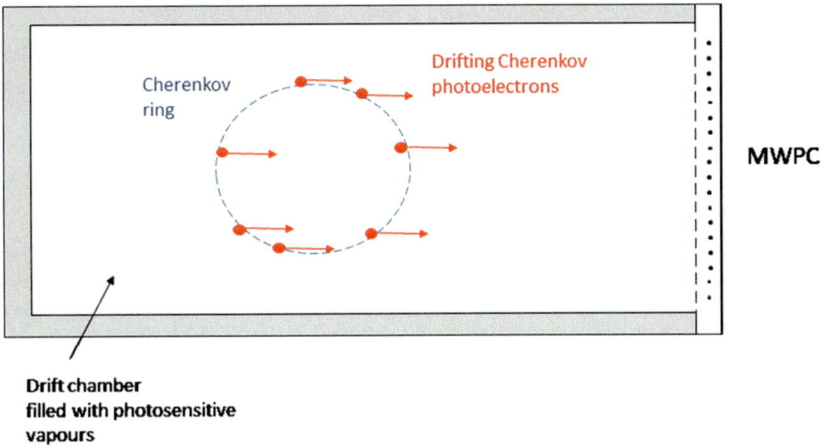

Figure 6-6. Top view schematic of the DELPHI RICH drift chamber, corresponding to the previous figure.

6.2.2. CsI-based RICH

After careful laboratory studies of the operation of gaseous detectors combined with CsI photocathodes, (briefly described in the previous paragraph), Charpak realized the advantages of these new devices:

1. Such detectors have even at room temperature a high quantum efficiency versus the wavelength comparable to TMAE vapours at full light absorption conditions. Thus, there is no need to install an additional heating system, CsI is equally effective for RICH.
2. CsI can be exposed to air, which simplifies the assembly of the detector.
3. RICH with CsI photocathodes have a much better time resolution compared to TMAE-filled ones due to the almost zero jitter between the photoelectrons' production and their detection.

Excited by this, Georges submitted three proposals to the CERN scientific committee: [Charpak, 1990,1-25], [Charpak, 1991, 1-12], [Braem, 1991, 1-15].

Unfortunately, neither of them was approved, probably because at this time Charpak was already retired and had not yet received the Nobel Prize in Physics.

However, these efforts of Charpak were highly appreciated in the CERN community.

Just one example. In 1991 the spokesperson of the RD26 collaboration F. Piuz visited George's office (one of us-VP- was a witness of this conversation) and shared with him an idea to reorient his team on building CsI-based RICH instead of TMAE-based RICH. At that time, it seemed like a risky twist, however, Charpak endorsed it with enthusiasm.

This was a turning point in the development of photosensitive gas detectors!

Very soon the RD-26 collaboration focused on studying the CsI photocathodes for the well-defined project: CsI-based RICH. Although the technology developed earlier at CERN R&D laboratories (Charpak, Seguinot, Braem) and summarized in [Peskov, 1977, 48-51] was fully confirmed, however, the real challenge was to extend it to large-area photocathodes needed for building a RICH detector and to achieve the necessary reproducibility and long-term stability.

The progress was impressive. Already in 1993, the first proximity focusing CsI-RICH prototype, shown schematically in Fig. 6-7, was built and tested on the particle beam [Almedia, 1994, 216-222].

Basically, it consisted of a NaF radiator and an MWPC combined with a CsI photocathode. These two essential elements were separated by the gap (proximity gap) necessary to separate the Cherenkov photons over measurable ring.

The encouraging results [Almedia, 1994, 216-222] suggested to propose CsI- based RICH detector for the ALICE experiment at LHC (see, for example, [Di Mauro, 1994, 284-287].

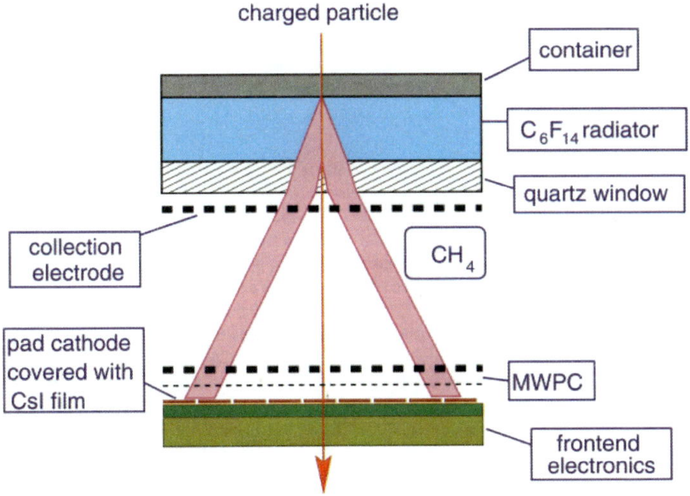

Figure 6-7. A sketch of a proximity focusing RICH prototype with a CsI photocathode (CERN image).

Following this, a special technology for mass production of large area (70x70 cm^2) CsI photocathodes was developed (see [Hoedlmoser, 2006, 338-350]).

The advancement in this area resulted in the construction of new RICH detectors for various other experiments, e.g., STAR, COMPASS and HADES.

6.3. MWPC for calorimeters

Another application of MWPCs is in calorimeters. In high-energy physics, calorimeters utilize a destructive method for measuring the energy of elementary particles. Layers of sufficiently dense material interleaved with detector layers are employed. Referring to Fig. 6-8, MWPCs serve as detector layers. A crossing particle or photon fully loses its energy, creating a shower of secondary particles and photons. As shown in Fig. 6-1, the electromagnetic calorimeter (transparent to hadrons but not to electrons and photons) precedes hadron calorimeters, where hadrons interact, producing showers of secondary particles.

In electromagnetic calorimeters, electrons and photons passing through matter primarily interact with the electron shells of atoms, generating a shower: a stream of numerous secondary electrons, positrons, and photons. Such showers develop and subsequently dissipate after a shallow depth of several tens of centimetres.

In a hadron calorimeter, high-energy hadrons (protons, neutrons, pi-mesons, and k-mesons) lose energy mainly due to collisions with nuclei. In this case, a hadron shower is produced, penetrating much deeper into the thickness of matter than an electromagnetic one and exhibiting different geometry, particularly larger transverse dimensions. Therefore, to fully absorb a hadronic shower from a particle of very high energy, a thickness of one or two meters of dense matter is required. Calorimeters not only measure energy but also determine the type of interaction: whether of electromagnetic or hadronic origin, which is crucial for a detailed understanding of the collision event.

For data analysis, it is important to determine the shape of showers and to obtain detailed 3D information about the deposited energy, requirements that are fulfilled by MWPCs arranged in a sandwich structure with absorbers as show in Fig.6-8.

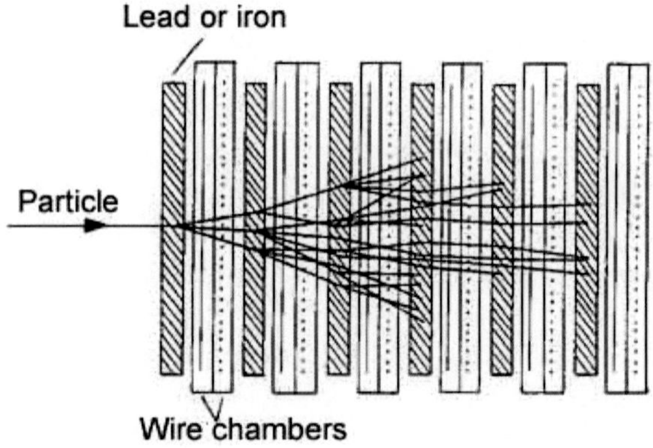

Figure 6-8. Schematics of a colorimeter with inserted MWPCs. CRRN image.

6.4. MWPCs in muon detectors

As seen in Fig. 6-1, the outermost layer of an experiment is equipped with muon chambers/trackers. A specific feature of muons is that they lose small amount of energy in the interaction with the matter bulk and, thus, can cross many meters of matter before stopping, penetrating where other particles practically cannot. There are two physical reasons for this:

1. Muons interact weakly with the matter and are relatively "heavy," therefore cannot effectively transfer energy to the electrons of atoms in a collision.
2. They do not participate in the strong interaction and are weakly scattered by nuclei.

This makes almost impossible to measure the energy of muons with calorimeters but allows for clear identification of muons from other particles thanks their detection in the outermost layers of experiments.

Two types of muon detectors are widely used in practice: scintillation and gas detectors. Scintillation detectors are typically made from suitable plastic. Gas detectors come in various designs: MWPC, drift chambers, and resistive plane-parallel chambers. To achieve high efficiency, two or more layers of trackers are employed (Fig. 6-9).

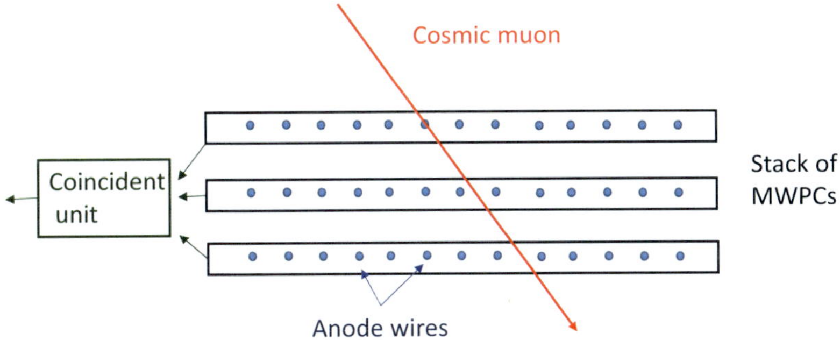

Figure 6-9. Schematic of a colorimeter with integrated MWPCs.

Usually, these detectors are interspersed with layers of iron yoke from a magnet. This enables the determination of the momentum of muons based on the trajectory bending within the magnetic field.

Concluding remarks

The flexibility of the MWPC to be built in sizes and performances tailored to different applications, such as TRDs, Cherenkov detectors, and calorimeters, highlights once again the remarkable versatility of this detector.

For the Curious: The Transition Radiation and Cherenkov radiation

Abstract

Transition radiation and Cherenkov radiation are primarily used for particle identification in experiments at accelerators, colliders, and in cosmic ray studies. The choice of radiator and photodetector determines the range of particle momentum over which identification is effective

The Transition radiation

Transition radiation was predicted by V. L. Ginzburg and I. M. Frank in 1945 [Ginsburg, 1945, 353-362]. According to calculations, it should occur when a charged particle crosses the boundary separating two media with different dielectric constants ε (Fig. 6-10). There is no threshold for the occurrence of this radiation; it is generated by any charged particle of arbitrary mass with a speed $v \neq 0$ in a system with low absorption of an electromagnetic signal.

Transition radiation was not studied experimentally for a long time, as the expected number of photons per interface was extremely small. The first measurements were taken in the optical range [Goldsmith, 1959, 836-844]. Interest in transition radiation increased sharply when the Soviet scientist G. Garibyan ([Garibian, 1958,1079-1075], [Garibian, 1960, 372-376]) showed that for ultra relativistic particles, the transition radiation should also appear in the X-ray range, which opened up new possibilities in experimental technology.

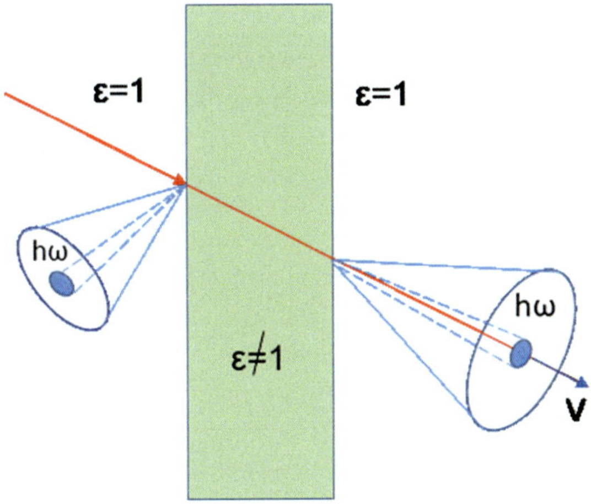

Figure 6-10. Schematic representation of transition radiation ($\omega = 2\pi\nu$).

As follows from the calculations, for a relativistic particle the intensity of the transition radiation in a wide energy range is proportional to the square of the charge of the particle $(Ze)^2$ and its Lorentz factor

$$\gamma = [1 - (v/c)^2]^{-1/2} \qquad (6\text{-}1),$$

where v and c are the speeds of the particle and light, respectively. The main part of the transition radiation is in the X-ray range of the spectrum and is concentrated inside a cone with an angle

$$\theta = 1/\gamma \qquad (6\text{-}2)$$

relative to the direction of particle motion. With reference to Fig. 6-10, the radiation cone originating from the front surface of the target is directed backward, making an angle of $\pi-\theta$, relative to the target, whereas the radiation cone from the back surface is directed forward, along the particle velocity (e.g., for an electron).

These properties of the transition radiation are used in experimental techniques to determine the charge and mass of particles at very high energies (>100 GeV) when other methods become ineffective.

Particles with the same energy but with different mass have different relativistic γ factor. Lighter particles for which $\gamma > 10^3$, can produce transition radiation, whereas heavier particles ($\gamma < 10^3$) do not. By detecting the TR, it is therefore possible to distinguish between these particles.

For instance, a substantial difference in masses enables the distinction of electrons from pions in the energy range from 0.5 GeV to 200 GeV, where $\gamma_{\mathrm{electron}} \approx 10^3$ already at 0.5 GeV, and $\gamma_{\mathrm{pion}} \approx 10^3$ only at 200 GeV. Since the probability of emitting a photon of transition radiation upon a single crossing of a medium boundary is small (1/100), layered or porous radiators with a low atomic number and many interfaces (up to several hundred) are employed. Typically, they consist of a stack of several hundred thin (5-100 microns) polypropylene films, regularly spaced, with a distance between the foils of ~ 0.1-2 mm.

Porous radiators are also used and may include polypropylene or carbon fibers, foam, or granular lithium hydride. These irregular radiators are much easier to assemble, even though they are slightly less effective at producing the TR than a regularly spaced radiators with the same total thickness.

Of course, it is essential that the radiator not only be effective at generating transition radiation but also absorbs it only weakly.

Candidates for detecting TR photons include proportional gas chambers, as well as cameras sensitive to visible light (see for example [Bhadra, 2013, 54-58].

How Cherenkov radiation was discovered

In the early years of the twentieth century, numerous researchers explored the enigmatic phenomenon of luminescence, in which photons are emitted without the generation of heat. It was observed that there are various types of such radiation; for example, they may be triggered by light, electrons, or radiation from recently identified radioactive substances. In the early 1930s, the Russian physicist Sergey Vavilov, who was deeply involved in luminescence research, suggested to his student Pavel Cherenkov that he investigate the luminescence produced by gamma rays in various materials dissolved in water.

A notable feature of these measurements was that the luminescence was so faint that Cherenkov had to remain in darkness for several hours to adapt his eyes before collecting data. After numerous frustrating attempts, he finally detected a faint blue glow emanating from the solvents. At first, he attributed it to contamination of the liquids. Indeed, earlier experiments with X-rays had suggested that luminescence in liquids occurred only in the presence of impurities.

Cherenkov meticulously redistilled several samples of water; yet they continued to exhibit faint luminescence when exposed to gamma rays. Together with Vavilov, he concluded that it might represent a novel effect. Nevertheless, nearby physicists frequently joked about the supposed delusions of an eager young student who had spent too long in darkness. Some, willing to accept his findings, dismissed them as ordinary fluorescence. Even the renowned Lev Landau counted among Cherenkov's sceptics. The scepticism deepened when *Nature* refused to publish Cherenkov's paper outlining his results, marking a significant setback. Undeterred, those convinced of the effect's uniqueness submitted the same paper to *Physical Review*, where it was ultimately accepted.

The international community held Cherenkov's discovery in high regard. In 1958, he was awarded the Nobel Prize in Physics alongside I. Frank and I. Tamm, who had elucidated the characteristics of this radiation. Notably, they were the first Russian physicists to receive this prestigious honor (Cherenkov's mentor, Vavilov, had passed away in 1951). Nevertheless, it took some time for his Russian colleagues to fully recognize Cherenkov's significant contributions to physics. Six years later, he was elected a corresponding member of the USSR Academy of Sciences, and six years after that, he became a full member.

We can only speculate why Cherenkov's Russian colleagues were initially so hesitant. In this book, we aim to explain how a brilliant concept revolutionized the field of particle and photon detectors. Another fascinating story could be told about how simple, careful observations can lead to the discovery of a new effect. Indeed, Cherenkov's observations were remarkably straightforward: he detected the light with his own eyes, without specialized instruments or complex techniques. Yet, it was the combination of these observations with the intellectual rigor of Cherenkov and Vavilov that led to their groundbreaking achievement. This is somewhat unusual in experimental physics, where discoveries are often the result of highly intricate and elaborate experiments, both then and now.

Underlying Physics of Cherenkov Radiation

As illustrated in Fig. 6-11, when a charged particle travels through a medium at a speed exceeding that of light in the material, it emits Cherenkov radiation. This condition is expressed as:

$$v > \frac{c}{n} \tag{6-3}$$

where c is the speed of light in vacuum, and n the refractive index of the medium.

An interesting feature of Cherenkov radiation is that the emitted light forms a cone with an angle θ, which is given by:

$$\cos\theta = 1/\beta n \qquad (6\text{-}4),$$

where $\beta = v/c$ is the particle's speed as a fraction of the speed of light in vacuum.

The particle generates a polarization wave in the medium, analogous to the shock wave produced by a supersonic jet in the atmosphere. In the simple case of photon detection on a plane perpendicular to the particle's trajectory, the Cherenkov photons form a circular pattern. The particle's speed can then be determined by measuring the emission angle of the photons, using the ring radius and the geometry of the detector.

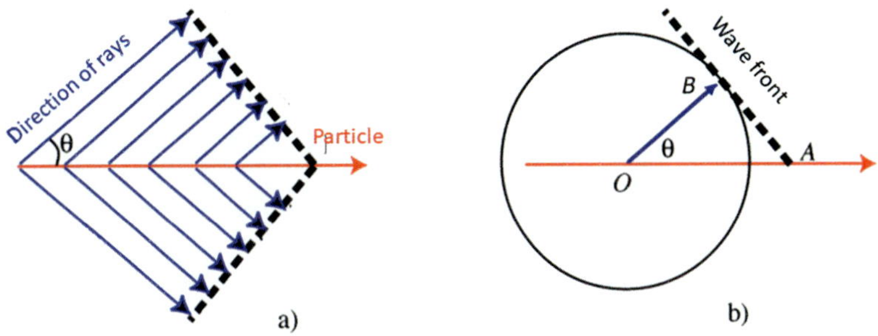

Figure 6-11. (a) *A coherent wavefront is formed at a specific angle θ (the Cherenkov angle) relative to the particle's trajectory when the particle moves faster than the phase velocity of light in the medium.* (b) *Geometry of the Cherenkov wave.*

The energy spectrum of Cherenkov light is continuous and is given by:

$$d^2N/dxd\lambda = 2\pi\alpha[1-1/\beta^2 n(\lambda)^2]/(\lambda)^2 \quad (6\text{-}4),$$

where N is the number of photons emitted per unit path length per unit wavelength, $\alpha \approx 1/137$ is the fine-structure constant, λ is wavelength of emitted radiation, $n(\lambda)$ is the wavelength-depended refractive index, x is path of the particle.

This formula shows that more photons are emitted at shorter wavelengths, especially in the ultraviolet and blue regions of the spectrum.

If an independent measurement of the particle's momentum ***p*** and electric charge sign is available, the relativistic formula can be used to determine the particle's mass:

$$m^2 = p^2(n^2 cos^2\theta_{Cherenkov} - 1) \quad (6\text{-}5)$$

The combination of the particle's charge and mass uniquely identifies the particle.

As described earlier, this technique of particle identification is widely used in particle physics experiments.

CHAPTER 7

MWPC APPLICATIONS IN HIGH-ENERGY PHYSICS EXPERIMENTS

Abstract

This chapter reviews the applications of wire-type detectors in various high-energy physics and astrophysics experiments, emphasizing those that were pioneers in utilizing groundbreaking technologies. Experiments that led to the most significant discoveries, such as the charm quark, the W and Z^0 bosons, the top quarks, and the Higgs boson, are also highlighted.

Introduction

In the previous two chapters, we described Charpak's invention of the MWPC and highlighted his pivotal role in the development of its successors, including drift chambers, TPCs, and RICH detectors. As is often the case, major discoveries and inventions emerge at a historical moment when the relevant science and technology are sufficiently advanced. Charpak's innovations are a prime example of this phenomenon.

In the early 1960s, the rapid development of particle accelerators created an urgent need in experimental high-energy physics for fast imaging tools capable of operating in high-flux particle and photon environments.

The analysis of these images allowed researchers to study the physics of particle and photon interactions, their decay pathways, and various transitions. However, the imaging detectors available at the time, such as Wilson chambers, cameras, and spark chambers, were inadequate for operation under high-flux particle conditions.

Charpak's MWPCs offered significantly higher speed and time resolution, along with self-triggered functionality, making it unsurprising that they were rapidly adopted. It was not unexpected, thus, that the Charpak chamber quickly (just a few years after their invention!) discovered extensive uses in high-energy physics experiments: in drift chambers, TPCs, muon detectors, and calorimeters.

In the following Section 7.1, we will outline several earlier experiments in which wire-type detectors played a significant role. Nevertheless, the primary focus of this chapter is on the use of MWPCs in modern high-energy physics experiments at CERN and in the United States.

7.1. Review of MWPC applications in high-energy physics experiments (1975-1989)

One of the initial uses of MWPC was in the WA-1 at CERN and E288 experiments at Fermilab, Batavia, Il, USA.

WA-1 experiment

WA-1(West Area-1) neutrino experiment which collected data from 1976 to 1984) aimed to study deep inelastic neutrino interactions in iron [https://cds.cern.ch/record/594735/files/CM-P00040240.pdf]. The detector was composed of 19 (later 20) magnetized iron modules, with drift chambers for track reconstruction and plastic scintillators positioned in the gaps between these modules. Each iron module served as the target, where interacting neutrinos produced hadron showers.

E288 experiment

In (1977) the experiment was installed in the proton fixed-target area of the laboratory under the direction of Leon Lederman, involving researchers from Columbia University, Fermilab and Stony Brook University [https://lib-extopc.kek.jp/preprints/PDF/1977/7708/7708130.pdf]. The detector integrated eleven MWPCs as muon tracking system for the momentum measurement; seven scintillation counter hodoscopes that determined muon trajectories; a drift chamber; a gas-filled threshold Cherenkov counter to reject potential low-momentum muon triggers; and an iron magnet to select the particle momentum.

This experiment reached its conclusion in 1977 [https://lss.fnal.gov/archive/1997/conf/Conf-97-432-E.pdf].

Through the analysis of muon tracking data and backtracking, a new particle, the Upsilon (Υ), along with its mass, was discovered. Further study of this particle's properties revealed that it consisted of a previously unknown type of quark–antiquark pair. The quark was formally named "beauty" but is more commonly referred to as the bottom quark. At the time, it was the most massive elementary particle known, with a mass of approximately $5\,\mathrm{GeV}/c^2$, more than five times that of the proton (see [https://digital.library.unt.edu/ark:/67531/metadc691485/m1/3/].

UA1 and UA2 experiments

Another remarkable physics result from this period, made possible by the MWPC's descendant, the drift chamber, came from the UA1 and UA2 experiments at CERN (UA stands for Underground Area), which operated from 1981 to 1990. As its central detector, UA1 employed a cylindrical drift chamber, a precursor of the TPC.
[https://cds.cern.ch/record/2103277/files/9789814644150_0006.pdf?subformat=pdfa&version=1].

Both experiments searched for W and Z bosons by identifying their decay products produced in proton–antiproton ($p\bar{p}$) collisions at the Super Proton–Antiproton Synchrotron (Sp$\bar{\text{p}}$S) facility. Detailed information about these significant discoveries can be found in Chapter 10, which we strongly recommend the reader consult.

In 1984, Carlo Rubbia and Simon van der Meer were awarded the Nobel Prize in Physics "for their decisive contributions to the large project, which led to the discovery of the vector bosons W and Z, mediators of weak interaction."

PEP-4 experiment

As mentioned in Chapter 4, PEP-4 was a pioneering experiment that, for the first time, employed a Time Projection Chamber (TPC). The PEP-4 facility (positron–electron–proton) was a large detector complex designed to measure and identify particles produced in electron–positron collisions with a centre-of-mass energy of 29 GeV at the Stanford Linear Accelerator Center's e^+e^- storage ring, PEP. It began collecting data in 1983.

This first large-scale TPC featured a cylindrical drift volume measuring two meters in length and two meters in diameter, filled with a gas mixture of 80% argon and 20% methane at 8.5 atm, presenting a significant technical challenge at the time. The TPC was positioned inside a solenoid and centred on the interaction point. A negative high voltage applied to the central membrane forced electrons to drift along electric and magnetic fields coaxial to the beam direction toward the two readout planes, one at each end (drift length: 1 m; see Figs. 5-6 and 5-8).

Each endcap housed MWPCs arranged in six sectors, each with azimuthal anode wires. This TPC measured ionization energy loss (dE_I/dx) versus particle momentum in multihadron events resulting from electron–positron annihilation. The success of PEP-4 prompted the rapid adoption of the TPC concept in other experiments. In fact, even while the PEP-4 detector was still under construction and its performance uncertain, physicists were already developing other time-projection chambers for experiments with simpler requirements (see, for example, [Madaras, 1984, pp. 36–47], and references therein).

OMEGA spectrometer

The Omega spectrometer was constructed in the early 1970s as a versatile facility, capable of operating under various triggers and with different incident beams [*The CERN Omega Spectrometer: 25 Years of Physics*, Proceedings, Editors: M. Jacob, E. Quercigh, CERN 97-02, 12 June 1997. https://cds.cern.ch/record/330556?ln=en.]. Its initial goal was to replace bubble chambers with electronic detectors, specifically spark chambers. Seven years later, the spark chambers were themselves replaced by MWPCs. The experiment also incorporated Cherenkov counters and large photon detectors. Cherenkov photons produced in a 5-meter-long radiator were detected using an array of 16 RICH detectors containing TMAE-based gas mixtures.

Research with the upgraded Omega spectrometer initially focused on the production of the charm quark, followed by studies of beauty quark production and searches for glueballs. The collaboration also developed and implemented a robust heavy-ion program aimed at producing particles containing strange quarks.

Up until 1988, the RICH detector participated in experiments WA69 and WA82. It was subsequently upgraded with new mirrors and drift-type chambers and employed in the WA89 and WA94 experiments. The Omega spectrometer thus demonstrated exceptional adaptability as a research facility.

These and other early efforts underscored the potential of MWPCs in high-energy physics experiments. Starting in 1980, the use of this detector began to attract significant interest among experimental physicists.

MARK I

The Mark I experiment was a pioneering particle physics project at the Stanford Linear Accelerator Center (SLAC), operating at the Stanford Positron–Electron Asymmetric Ring (SPEAR) collider in the early 1970s. It was the first detector to provide uniform coverage over the full 4π solid angle around the interaction point, with electrons and positrons colliding within a layered arrangement of diverse detector types.

MWPCs were employed inside a solenoidal magnet, alongside spark counters, to measure the momenta and trajectories of charged particles (see, for example, Feller et al., 1977, pp. 1–5).

This design proved highly successful and led to the discoveries of the J/ψ meson and the τ lepton, discoveries for which Burton Richter (1976) and Martin Lewis Perl (1995) were awarded the Nobel Prize in Physics. The fundamental design principles established by Mark I continue to influence modern collider detectors.

MARK II

The Mark II detector was installed in 1977 at the SPAIR storage rings at SLAC (an e+e⁻ collider in the 3–5 GeV energy range) to study electron-positron annihilations. In 1979, it was moved to PEP (an e+e⁻ collider at 29 GeV), where the research program was expanded to include searches for top quarks, jets, gluons, B mesons, etc. Originally, it had 16 layers of cylindrical drift chambers surrounding the beam pipe to measure the azimuthal coordinates of charged tracks with an RMS (root mean square) accuracy of 200 μm at each layer [Abrams ,1989, 55-80].

In 1982, Mark II was selected to be the first detector at SLC (an e+e⁻ linear collider at the Z pole/the resonance peak - the energy level at which the Z boson is produced in large quantities).

In 1985, the detector was substantially upgraded. A new central drift chamber was built to:

a) achieve good momentum resolution in the existing 5.5 kG field,
b) preserve good solid angle coverage,
c) improve pattern recognition and high tracking efficiency related to Z boson physics events,
d) exploit dE_l/dx as an independent aid to calorimetry for electron-hadron separation for momenta less than about 10 GeV/c.

Afterwards, Mark II was returned to PEP for testing. In 1987, it was rolled onto the SLD (SLAC Large Detector) beamline and took data at 91 GeV, performing, for the first time, an energy scan of the Z_0 resonance. In 1988, a vertex drift chamber was developed for Mark II to be combined with the central tracking chamber [Alexander, 1989, 519-527]. In 1990, Mark II was replaced by the SLD detector.

MARK-J

The physics objectives of the MARK J collaboration at PETRA (the electron-positron colliding beam accelerator at DESY in Hamburg, which operated from 1978 to 1984) were studies of several purely electromagnetic processes and hadronic final states at high-energy electron-positron collisions. It provided a particularly clean way to investigate the nature of the fundamental constituents of matter and their interactions because of the point-like nature of leptons (see for example, [Adeva, 1984, 131-226]). Experiments at PETRA were thus free of the complications that arise in studies of hadron-hadron or lepton-hadron collisions, where the hadrons themselves have a complex internal structure. Correspondingly, the MARK J detector was designed to distinguish charged hadrons, electrons, muons, neutral hadrons, and photons and to measure their directions and energies. Besides other detectors, it contained a drift tube-based vertex detector and a system of wire chambers: an inner drift chamber of a muon spectrometer and drift chambers inside and outside the magnet for momentum measurements [Anderhub, 1988, 50-59].

HELIOS/NA34 Experiment

HELIOS/NA34 (1984–1988) was a fixed-target experiment conducted at the SPS facility at CERN. It was designed to address open questions regarding the production of soft photons, electrons, muons, and neutrinos in hadron interactions [see, for example, J. Schukraft, "Recent Results from HELIOS (NA34) on Proton-Nucleus and Nucleus-Nucleus Reactions," CERN-EP/88-176, 15 December 1988, https://cds.cern.ch/record/193646/files/198901259.pdf].

This multipurpose apparatus included silicon strip trackers, transition radiation detectors, uranium/liquid scintillators, and uranium-liquid argon calorimeters. A notable technological feature was the use of high-resolution drift chambers for charged particle momentum measurements in a dipole magnet, as well as an MWPC-based Xe-filled TRD detector. These additions significantly enhanced the capabilities of the HELIOS spectrometer.

CDF Experiment

The Collider Detector at Fermilab (CDF) is a calorimetric detector that covers nearly the entire solid angle around the interaction region [Abe, 1988, 387-403. The volume inside the magnet is occupied by a large cylindrical drift chamber and a set of Time Projection Chambers. Muon detection is performed with drift chambers outside the calorimeters in the central region, and with large, magnetized steel toroids and associated drift chambers in the forward and backward regions.

Among Fermilab's physics programs was the search for the top quark. The early discovery of the bottom quark strongly suggested the existence of a sixth quark, the top, to complete the quark pair. It was known that this quark would be heavier than the bottom quark, requiring more energy for its production in particle collisions.

DØ detector

Until the start of the Large Hadron Collider operation at CERN in 2009, the Tevatron was the only collider powerful enough to produce top quarks. To confirm the expected discovery, a second detector, DØ (Detector at Zero Degree), was added to the Fermilab accelerator complex. Its name reflects its placement at one of the interaction points along the accelerator ring, specifically at zero degrees with respect to the proton beamline.

The primary physics focus of the DØ experiment was the study of high-mass states and large transverse momentum (p_t) phenomena [Abachi, 1994,185-253]. In addition to the search for the top quark, these studies included precision measurements of the W and Z bosons, which provided sensitive tests of the standard electroweak model, various studies of perturbative Quantum Chromodynamics (QCD), the production of b-quark hadrons, and searches for new phenomena beyond the Standard Model. The DØ detector contained several drift chambers: the transition radiation detector, the central drift chamber, and two forward drift chambers. The central drift chamber provided coverage for tracks at large angles, after the transition radiation detector and just before entering the central calorimeter.

The detector also contained drift tubes, which were essentially drift chambers assembled from separate single-wire modules. The principle of their operation was based on Charpak's idea of forming a uniform electric field in a single-wire cell by biasing cathode strips with linearly increasing voltages (see Fig. 5-5).

In October 1992, the first hint of the top quark was found. In the following years, more evidence was collected, and both experiments ultimately confirmed the existence of the top quark (see, for example, [Riordan, 1992, 1-36]).

As a culmination, on February 24, 1995, the DØ community submitted a paper to Physical Review Letters declaring the discovery of the top quark, exactly at the same time as the CDF team.

ARGUS (A Russian-German-United States-Swedish Collaboration (~1982-1992); an experiment at the electron-positron collider in the DESY ring DORIS II. [Albrecht, 1989, 1-48]. It aimed to explore the properties of the charm and bottom quarks. The ARGUS was a hermetic detector and had 90% coverage of the full solid angle. It had drift chambers, a time-of-flight system, electromagnetic calorimeters, and muon chambers (proportional drift tubes). It was the first experiment that observed the mixing of the B mesons.

SLD detector

The SLD was a general-purpose apparatus designed to study e+e− interactions at energies up to 100 GeV in the center of mass at the Stanford Linear Collider (SLC). Its design was optimized for physics at the Z_0 resonance. To achieve this, several new technologies were implemented, such as the CCD vertex detector [SLD Design Report, SLAC Report 273 (revised 1985) https://inspirehep.net/files/9cd2471c8c8050783573f210f09cf780]).

MWPCs were used in a high-precision central drift chamber (see for example [Young, 1986, 176-177], pairs of endcap drift chambers, and a RICH detector. The latter, named CRID (Cherenkov Ring Imaging Detector), was the most notable wire-based detector of its time. The early R&D developments of the Barrel CRID are summarized in [Leith, 1988, 120-136]. Its design was similar to that of the DELPHI RICH (see previous chapter), and each group benefited from the work of the other. In these new detectors, Cherenkov photons emitted in a cone around the charged-particle trajectory were focused onto an array of 40 TPCs. The three-dimensional location of each photoelectron conversion was recorded in the TPCs, and the production angle relative to a given charged track was reconstructed. As mentioned in the previous chapter, the use of TMAE in gaseous detectors presented a great challenge due to its chemical aggressiveness. Solutions to some of the associated problems were found during R&D activities in Charpak and Seguinot's labs. However, in a large TPC, many other difficulties arose, such as gas chamber leaks, the lifetime of drifting electrons, wire aging, self-sustaining cathode current, and others. The SLD team invested considerable effort to resolve these issues. As a result, one of the major achievements of the CRID was to convince, despite the engineering challenges, many skeptics in the high-energy physics community that the RICH concept could be implemented in practice. This was later confirmed by the DELPHI experiment.

Experiments at TRISTAN

Since the Stanford Linear Collider (SLAC) and the Large Electron-Positron Collider (LEP), which began operation in 1989, the TRISTAN e^+e^- collider at the High Energy Accelerator Research Organization in Japan (KEK) had ceased its role as an energy-frontier machine for several years. The emphasis shifted toward high-statistics measurements at center-of-mass energies of around 60 GeV. TRISTAN was not a typical ring accelerator; in fact, it resembled four linear accelerators connected by four quadrants of arcs with a relatively short bending radius.

There were four colliding sections, each with a corresponding detector. The acronyms of these detectors were: AMY, SHIP, TOPAZ, and VENUS. SHIP was a small detector for heavy ionizing particles with plastic sheets, while the other detectors had designs closer to classical ones and contained relatively large multiwire-type detectors. For example, the TOPAZ TPC had a diameter of 2.6 m and a length of 3 m, allowing for three-dimensional particle tracking and particle identification due to dE_l/dx measurements. This capability was essential for many studies, including those involving low transverse momentum and charm production, where electron identification provides strong evidence for resolved-photon processes, as well as for higher-order corrections in QCD.

VENUS and AMY also featured cylindrical central drift chambers, but with smaller diameters. Both TOPAZ and VENUS had cylindrical inner drift chambers. Some physics results obtained with these four detectors can be found, for example, in [Takasaki, 1988, 17-37].

7.2. Experiments in the LEP era (1989-2000)

LEP

The renewed momentum in the development and application of wire-type detectors contributed significantly to LEP (Large Electron-Positron Collider) and other experiments of the era.

LEP, the largest collider for electrons and positrons, operated from 1989 to 2000. It went through two phases: from 1989 to 1995, it focused on the production of Z bosons (the neutral carriers of the weak force), while from 1996 to 2000, it studied W bosons (the charged carriers).

At LEP, the beams collided at four points, equally distributed over the 27 km circumference of the ring. Large detectors, ALEPH, DELPHI, L3, and OPAL, were installed at each collision point in vast underground caverns, ranging from 80 to 140 meters deep.

Their primary scientific goals were to determine the precise masses of the W and Z bosons and to confirm the existence of exactly three light neutrino species. The same beams crossed at multiple interaction points, ensuring that these experiments collected data at the same energies and with roughly the same event rates, while maintaining a healthy competition to achieve results.

ALEPH (Apparatus for LEP Physics) was an advanced-technology, general-purpose detector (see, for example, [Decamp, 1990, 121-178], Buskulis, 1995, 481-506]). The main feature of ALEPH was a time-projection chamber, larger than the one developed earlier at the Lawrence Berkeley Lab and installed in the PEP e^+e^- collider at SLAC. In addition, in ALEPH multiwire detectors were also used in the drift chamber surrounding the vertex detectors (Inner Tracking System) and in the electromagnetic calorimeter (EM) for shower detection. The EM consisted of a 45-layers lead/MWPC sandwich, with a total thickness of approximately 20 radiation lengths.

DELPHI (Detector with Lepton, Photon, and Hadron Identification) was a 4π solid angle detector with an emphasis on particle identification, three-dimensional information, high granularity, and precise vertex determination [Aarino, 1991, 233-276]. It featured several MWPC-based detector systems, such as the inner tracker, TPC, barrel muon chamber, high-density projection chamber, and an array of drift tubes, including in the outer detector and barrel muon chambers. A unique feature of DELPHI was its drift chamber-based Ring Imaging Cherenkov detector, which measured particle velocities and enabled the identification of secondary charged particles. The RICH system consisted of three components: two forward RICH detectors and the barrel RICH detector. For details, see Figs. 6-5 and 6-6, and the description in the preceding chapter.

We briefly recall here that the forward DELPHI RICH detector employed two types of radiators: a liquid radiator (used in the proximity-focused RICH) and a gas radiator, coupled with a system of focusing mirrors. The resulting Cherenkov light cone was intercepted by a TPC-type photodetector.

In the barrel RICH, a liquid radiator was used. As in the forward RICH, the Cherenkov light was detected by a TPC photodetector, operating on the proximity-focusing principle.

The L3 detector consisted of three wire-type chambers: a central tracker detector, a hadron calorimeter, and a muon spectrometer [[Adeva, 1990, 35-102]. The central tracker, known as a time expansion chamber (see paragraph 5.1 in Chapter 5), was a modified and enlarged version of the drift chamber developed in the study by T. Bressani, et al [Bressani, 1969, 275-281].

The uranium hadron calorimeter consisted of two sections: the barrel and the forward-backward. The barrel section was a fine sampling calorimeter made of depleted uranium absorber plates interspersed with proportional wire chambers. The endcaps consisted of stainless-steel containers filled with alternating layers of brass tube proportional chambers

The muon detector was designed to measure high-energy muons with an accuracy of $\Delta p/p = 2\%$ at a momentum of $p = 50$ GeV/c. This performance was achieved using a configuration of three layers of drift chambers that precisely measured the curvature of the muon trajectory in the region between the support tube and the magnet coil.

OPAL (Omni-Purpose Apparatus for LEP) was a large, general-purpose detector that extensively used wire-type detectors [Ahmet, 1991, 275-319].

In this apparatus, the vertex drift chamber and the silicon tracker operated as a unified system to precisely locate the vertices of short-lived particles and improve the momentum resolution. The central jet chamber also featured a wire-based design. It could record events with high local track density (jets), offering excellent space and double-track resolution, as well as the ability to perform particle identification within a solid angle close to 4π steradians. This was crucial for particle identification via energy loss measurement (dE_l/dx) and for enhancing spatial resolution.

This system efficiently identified tracks in the plane perpendicular to the beam axis and was complemented by the so-called "z-chambers." These drift chambers were mounted around the outer mechanical support of the jet chamber and provided precise measurement of the z-coordinate (along the beam axis) of the tracks, improving invariant mass resolution.

MWPCs were used in the endcap electromagnetic calorimeter and in the pole-tip hadron calorimeter. The barrel muon detector contained approximately a hundred drift chambers. Additionally, two planes (forward and rear) of the drift chamber were installed in the forward detector to measure the luminosity of LEP.

After eleven years of operation, LEP was switched off in November 2000. A month later, dismantling began to make way for its successor, the Large Hadron Collider.

The widespread and highly successful use of MWPCs in experiments brought Charpak significant fame, ultimately leading to his Nobel Prize in Physics in 1992.

While the LHC was under construction, the leadership in heavy ion physics was held by the RHIC (Relativistic Heavy Ion Collider), located at Brookhaven National Laboratory in Upton, New York.

RHIC

RHIC has been in operation since 2000 and was the world's first collider capable of accelerating and colliding heavy ions at relativistic speeds, enabling physicists to study the primordial form of matter that existed shortly after the Big Bang. By colliding spin-polarized protons, RHIC also investigates the spin structure of the proton. Initially, it hosted two small experiments, BRAHMS and PHOBOS, and two large ones: PHENIX and STAR. The smaller experiments were later decommissioned, while PHENIX and STAR remain operational.

In *BRAHMS* (Broad Range Hadron Magnetic Spectrometers), wire-type detectors were used in the tracking system, which combined small TPCs with drift chambers.

PHENIX (Pioneering High Energy Nuclear Interaction Experiment) used various multi-wire technologies in its tracking system, including drift chambers, pad chambers, and a so-called time expansion chamber (see paragraph 5.1 in Chapter 5).

These sub-detectors were optimized for different functions. The low-mass multi-wire focusing drift chambers provided high-resolution transverse momentum (p_T) measurements necessary for determining the invariant mass of particle pairs. Three pad chambers provided three-dimensional space points for the charged tracks. The time expansion chamber identified tracks originating from outside the fiducial volume and provided electron-proton separation in the momentum range from 250 MeV/c to 2.5 GeV/c using dE_l/dx information.

The PHENIX drift chambers were cylindrically shaped and located in the region from 2 to 2.4 meters from the z-axis, extending 2 meters along the beam direction.

In combination with other detectors, it was possible to record various particles produced in RHIC collisions, such as photons, electrons, muons, and hadrons, although over a limited range of solid angles.

A summary of the physics results can be found in numerous papers, such as [Belmont, 2019, 22-28].

STAR (Solenoidal Tracker at RHIC) is a solenoidal apparatus focused on large solid-angle detection of hadrons. The heart of the detector is a 4π TPC situated in a magnetic field [Ackermann, 2003, 624-632]. The TPC chamber identifies and tracks particles emerging from heavy-ion collisions in the mid-rapidity region. The forward rapidity region is also covered by tracking using two end-cap forward TPCs. One of the main features of STAR is its ability to track a large number of particles produced in each ion collision.

To perform tracking in the forward region, a radial-drift TPC was added to the STAR detector system.

STAR was also the first collider experiment to use a relatively small RICH module based on MWPC, combined with a CsI photocathode, which extended the particle identification beyond the TPC and TOF ranges in a limited solid angle. This detector named 'Proto-2' was developed in collaboration with the ALICE RICH team.

7.3. Some other experiments

In the time gap between LEP and LHC, some other, albeit smaller-scale, experiments using MWPC were in operation. A detailed review of these is beyond the scope of this book; we will list only a few of them.

The **NA49** experiment (North Area, 1991-2002) investigated the properties of quark-gluon plasma using a large acceptance magnetic and tracking spectrometer. This spectrometer was designed to detect about 70% of all charged particles created in a single Pb-Pb collision event. The experiment used four TPCs: two in the vertex region and two in the forward region [Afanasiev, 1999, 210-244].

The **H1** experiment (1992–2007) [Abt, 1997, 348-396] at HERA (Hadron-Electron Ring Accelerator) facility aimed to investigate the internal structure of a proton through deep inelastic scattering measurements, as well as to study fundamental interactions between particles and search for particles beyond the Standard Model. The H1 detector's tracking system included multiwire jet (can detect many tracks at once) and Z chambers (designed to measure the Z coordinate (along the beam axis), as well as an array of trigger proportional chambers). One of the interesting features of the system was the combination of radial wire drift chambers with a multilayer transition radiator. This design allowed for simultaneous optimal track point measurements and efficient X-ray detection. As in many other experiments, the H1 muon system was equipped with modular drift chambers, a concept proposed in Charpak's work [Charpak, 1973, 413-426].

7.4. LHC era (2008-present time)

The Large Hadron Collider (LHC) is nowadays the world's largest particle accelerator, which is in operation since September 2008. The LHC consists of a 27-kilometer underground ring containing superconducting magnets combined with accelerating structures to boost the energy of the particles along the way.

There are several experimental stations located along the ring: three big experiments (ALICE, ATLAS, SMS), one medium size (LHCb)- see Fig. 7-1, and six relatively small (MilliQan, MATHUSLA, FASER, LHCf, MOEDAL, and TOTEM). Each is close to one of the larger experiments and uses the same collision point.

ATLAS and CMS are the latest examples of the extensive use of wire chambers.

ATLAS experiment

One of the main goals of the ATLAS experiment (A Toroidal LHC Apparatus) was the search for the Standard Model Higgs boson and this defined the performance of important sub-systems of ATLAS [Aad, 2008, 1-407]. ATLAS detector is nominally forward-backward symmetric with respect to the interaction point (see Fig.7-2).

In this detector, MWPCs were used in a large muon spectrometer which defines the overall dimensions of the ATLAS detector (around 44 m in length and 25 m in diameter). Two types of multiwire detector wire are used: the so-called cathode strip and thin gap chambers.

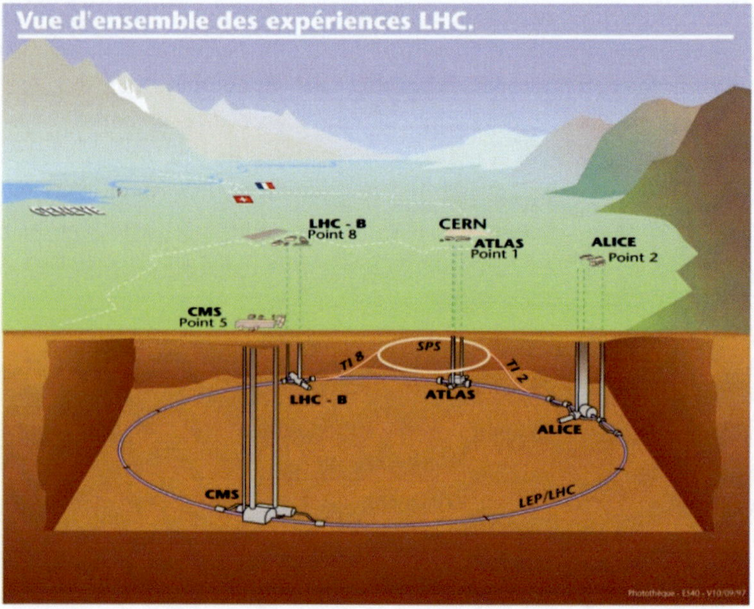

Figure 7-1. Layout of main underground CERN experiments and access points to them. Image credit: CERN.

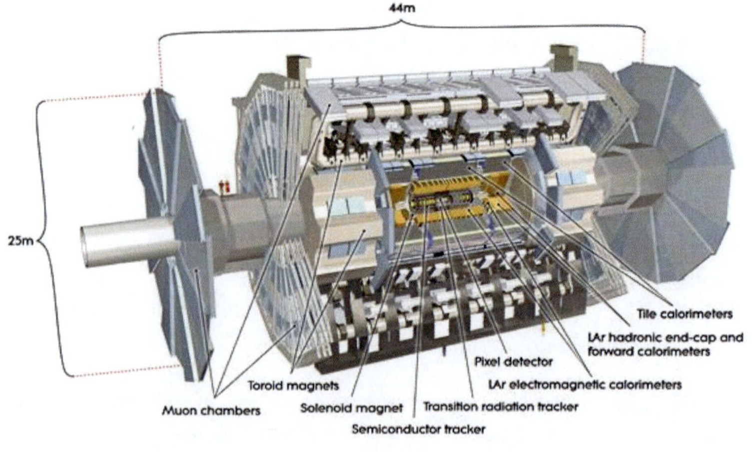

Figure 7-2 Schematics of the ATLAS experiment. Image credit: CERN.

The name "cathode strip" was given to the first detector because its cathode planes were segmented into strips in orthogonal directions. This design allows both coordinates to be measured from the induced charge distribution. These detectors are used in the innermost tracking layers of the forward region due to their higher rate capability and adequate time resolution. A thin gas chamber covering a much larger area was used in the ATLAS endcap. It was a slightly modified version of the OPAL thin-gap chamber, with a cathode coated by a graphite layer. As a first-level trigger detector, it had to provide high efficiency and excellent time resolution for bunch-crossing tagging in a high-background environment.

CMS experiment

The Compact Muon Solenoid (CMS) experiment at the LHC is another experiment designed to explore the fundamental building blocks of matter and the interactions that govern them. The search for the Higgs boson was a key objective of its program. A schematic of the detector is shown in Fig. 7-3.

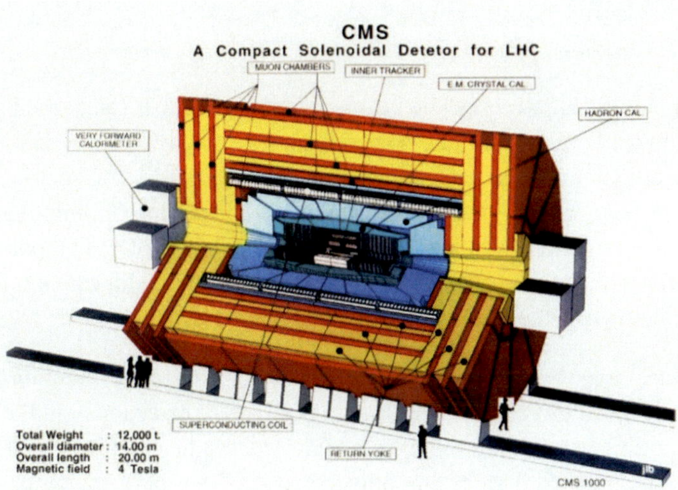

Figure 7-3. Artistic picture of the CMS experiment. CERN image.

Like the ATLAS experiment, CMS also uses cathode strip chambers as part of its detection system [Bayatian, 1994, 1-254]. These chambers are placed in the outermost layer of the CMS detector to detect and measure muons.

In 2012, the ATLAS and CMS collaborations announced the discovery of the Higgs boson, based on data collected during the first run of the LHC. This discovery marked a major milestone in particle physics, confirming the existence of the Higgs field, which gives particles their mass. MWPCs, together with other detectors, played a crucial role in this achievement: their high spatial resolution and fast response time enabled the detectors to reconstruct the tracks of charged particles and identify the decay products of the Higgs boson. This discovery has had a profound impact on our understanding of the fundamental laws of nature and will continue to shape the direction of particle physics research for years to come.

Among the smaller LHC experiments, it is worth mentioning the use of MWPCs in the LHCb muon system. With its 1,368 wire chambers, the LHCb muon detector is one of the largest instruments of its kind worldwide, and one of the most irradiated [Kotriakhova, 2020, 162787].

ALICE experiment

The aim of the ultra-relativistic heavy-ion program at CERN, carried out by ALICE (A Large Ion Collider Experiment), is to study strongly interacting matter at extreme densities and temperatures (T $\sim 10^{12}$ K; for comparison, the core temperature of the Sun is $\sim 10^{7}$ K). Quantum Chromodynamics predicts a transition from hadronic matter (ordinary matter) to a plasma of deconfined quarks and gluons, the fundamental building blocks of ordinary matter, known as the quark–gluon plasma (QGP) [Ahmad, 1995, 1-237].

This transition occurred in the early universe in the opposite direction, about 10^{-5} seconds after the Big Bang, and may still take place today in the cores of collapsing stars. Ultra-relativistic heavy-ion collisions provide the only means to access and study the universe at such an early stage, an epoch unreachable by astrophysics and cosmology, since the universe remained opaque to observation until about 380,000 years after the Big Bang. The ALICE experiment consists of two main components: a central detector, located inside a solenoidal magnet, and a forward muon spectrometer. An artistic rendering of the ALICE experiment is shown in Fig. 7-4.

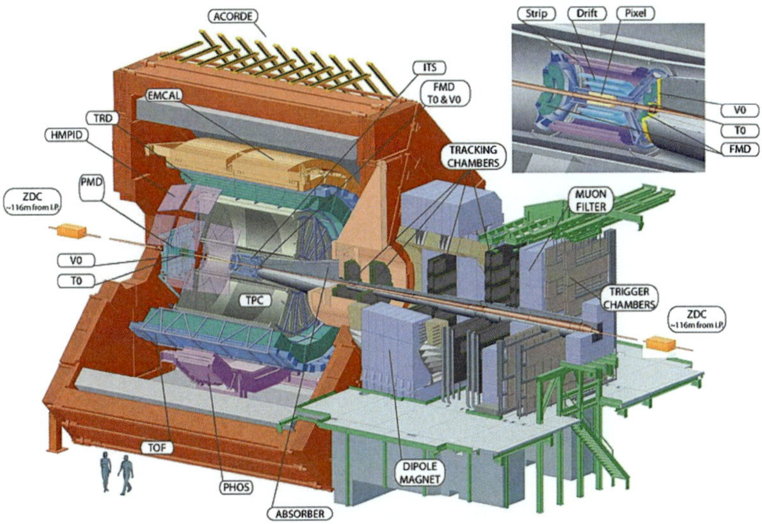

Figure 7-4. Artistic view of the ALICE detector. It composed from several detectors. MWPCs are used in TPC, RICH, TRD and in muon spectrometer. CERN image.

The ALICE TPC is the largest in the world, featuring an overall 'conventional' layout but with nearly all design parameters pushing beyond the state of the art. It occupies the radial region from 90 cm to 250 cm and is designed to achieve a rate-of-energy-loss resolution better than 7%. It also provides electron identification for momenta up to 2.5 GeV/c. Further details can be found in section 7.5.

Another notable wire detector is the ALICE RICH, which is part of the HMPID (High Momentum Particle Identification Detector) project. It is described in detail in the additional readings of this chapter. Before mass production, a prototype of the RICH module, named Proto-2, was built at CERN, shipped by air to the USA, and installed in the STAR experiment at Brookhaven's RHIC. The test results met the RICH design specifications, providing the experiment with good Particle identification (PID) capabilities despite the limited acceptance of Proto-2.

The HMPID is an important component of particle identification in ALICE. Conventional MWPCs were also used in ALICE in the Transition Radiation Detector and the muon system.

7.5. Concluding remarks

Based on the material presented in this chapter, it is evident that following the invention of the Multi-Wire Proportional Chamber, its successors, such as drift chambers, time projection chambers, and ring-imaging Cherenkov detectors, have been widely adopted in high-energy physics experiments. These detectors have proven highly effective in observing and analysing the properties of subatomic particles, providing valuable insights into the origins of the universe and the structure of fundamental particles.

Overall, Charpak's contributions to high-energy physics were profound. His inventions and methodologies revolutionized the field and paved the way for numerous significant discoveries. As a pioneer in the development of electronic imaging detectors, Charpak's work has left a lasting impact on the ways in which physicists explore the fundamental properties of matter.

7.6. Level Up: MWPCs in the ALICE experiment

Abstract

For readers with the appropriate background, this paragraph offers a detailed overview of the use of MWPCs in three ALICE experiment detectors: the High Momentum Particle Identification Detector (HMPID), the Time Projection Chamber (TPC), and the Transition Radiation Detector (TRD).

Introductory Note

Drawing on our direct involvement as members of the ALICE Collaboration, we provide an overview of the design, operation, and performance of MWPCs within this complex experimental environment. Our perspective combines firsthand experience with the broader context of detector development and implementation in high-energy physics.

7.6.1 The MWPC for the ALICE HMPID

As described in section 6.2.2, research at CERN in the early 1990s began exploring an alternative approach to building RICH detectors, using MWPCs with pad-segmented, CsI-coated photocathodes instead of TMAE.

The aim was also to develop a cost-effective solution compared to photomultiplier tubes (PMTs), suitable for large-area collider experiments. A summary of this work was published by François Piuz [Piuz, 2003, 76-90].

This R&D effort ultimately led to the successful construction of multiple RICH modules, with active areas of up to 10 m^2 [Beole, 1998, 1-188], achieved at a fraction of the cost. Covering the same area with PMTs would have cost approximately 18 million CHF; in contrast, the RICH modules equipped with CsI photosensitive MWPCs, offering a quantum efficiency of 25–30%, reduced the cost by a factor of ten. This represented a remarkable achievement.

HMPID was proposed in 1995 as part of the ALICE experiment [Ahmad, 1995, 1-237] at CERN's Large Hadron Collider. Following approval, the HMPID was installed in ALICE in 2006 (see Fig. 7-4). It was designed to extend the ALICE PID capabilities in separating pions, kaons and protons up to 5 GeV/c, while the Time Of Flight (TOF) detector was limited to 3 GeV/c. Fig. 6-7 illustrates the operational principles of the proximity-focusing RICH used in the HMPID. This module utilizes three Nobel Prize-winning processes: the photoelectric effect (A. Einstein, 1905), the Cherenkov effect (P. Cherenkov, 1937), and the MWPC-based detection (G. Charpak, 1968).

With reference to Fig. 6-7 Cherenkov photons produced in the liquid C_6F_{14} perfluoro-hexane radiator, upon passing through a fused silica window (which is transparent to UV Č photons), travel through methane (CH_4) until they are converted into photoelectrons on the pad-segmented, CsI-activated photocathode (reflective mode). The CH_4 flushing the MWPC provides the gas needed for avalanche multiplication near the anode wires. The Front-End Electronics (FEE) sense the signal on the pads.

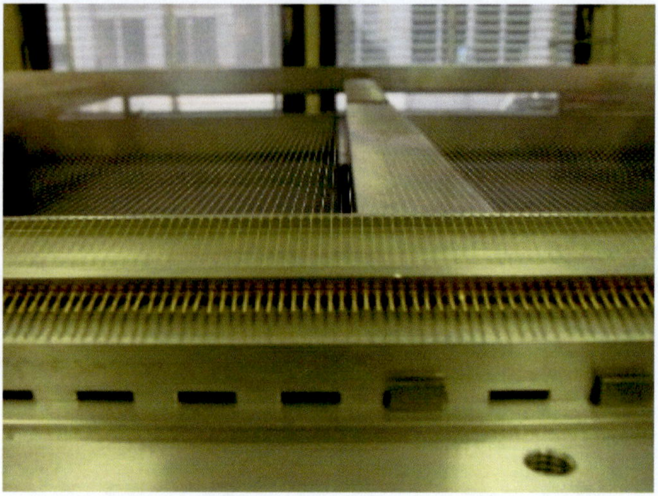

Figure 7-5. Photograph of the wire plane in the HMPID MWPC. From the HMPID photo archive of one of the authors.

Figure 7-6. Photograph of one of the six pad-segmented photocathodes per MWPC. From the HMPID photo archive of one of the authors.

Figs. 7-5 and 7-6 show photographs of the MWPC's anode wire plane and a CsI pad-segmented photocathode. Its production technique and quality evaluation are described in [Hoedlmoser, 2006, 338-350].

In this geometry, optimal Cherenkov imaging (via computer simulation) requires a pad area of approximately 1 cm^2, resulting in 2.3×10^4 channels/RICH module, and a total of roughly 1.6×10^5 FEE channels for the seven modules.

Fig. 7-7 shows the full HMPID with its seven RICH modules during transportation to the LHC experimental cavern UX25 before the installation in the ALICE experiment. The initial results of its successful operation are presented in several publications (see, for example, [Braem, 2007, 1144–1147]).

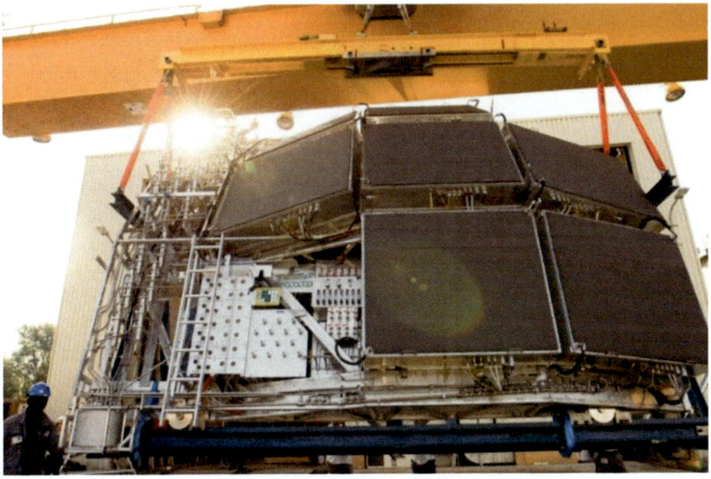

Figure 7-7. HMPID cradle during transportation to the ALICE experimental site, with the cooling panels of the protection box covered by slabs of insulating foam (September 2006). From the HMPID photo archive of one of the authors.

While there are many advantages in using a CsI, proximity focusing RICH detector, there are also factors that may restrict their use, particularly the charge dose on the CsI photocathode and the presence of contaminating gases in the MWPC, such as O_2 and H_2O. Indeed, beyond specific thresholds, a decrease in the CsI QE may occur. This can worsen the Cherenkov angle measurement and ultimately influence the PID performance [Braem, 2005,187-195], [Hoedlmoser, 2007, 28-38].

Fortunately, the observed event multiplicity of charged tracks produced in pp and Pb−Pb collisions at LHC energies, has been approximately six times lower than anticipated from computer simulations, thus reducing the charge load on the CsI photocathode. Moreover, rigorous monitoring of the gas quality supplying the MWPC has played a significant role in ensuring stable detector performance. This will allow the detector to continue functioning until 2026, marking the conclusion of the LHC Run 3 phase, even though the HMPID was initially expected to stop operating by the end of 2013. This represents a remarkable technological achievement, far exceeding initial expectations. In this instance, the RICH detector employing MWPCs with CsI-coated photocathodes has outperformed the more expensive PMTs, meeting all experimental requirements while delivering comparable physics performance. Similar detectors have been successfully employed in various other experiments, including COMPASS (Common Muon Proton Apparatus for Structure and Spectroscopy) at CERN's SPS, HADES (High Acceptance DiElectron Spectrometer) at GSI in Darmstadt, Germany, STAR at the RHIC collider in Brookhaven, USA, and JLAB-Hall A in West Virginia, USA [Nappi, 2005, 1-130].

Figure 7-8 shows the pad occupancy in the seven RICH modules of a Pb-Pb event at LHC. The pink curves represent the reconstructed Cherenkov patterns for primary particles originating from the collision vertex. Based on the results of a dedicated algorithm, Hough Transform (HT), to compute the Cherenkov angle for a given particle, circles and open-conic pink lines overlap the cluster map, corresponding to particles impinging on the CsI photocathode perpendicularly or at an angle. Visually, other patterns can be spotted; however, they correspond to secondary particles, which are not relevant for the charged hadron spectra that will be produced offline through the HMPID data analysis. In this event, with a high multiplicity of clusters in the RICH6 detector, the right half of the detector shows a reconstructed pattern, most likely using background clusters. Chapter 10 discusses the promising use of Artificial Intelligence (AI), which appears to be more efficient than the HT in removing these false events. Fig. 7-9 shows a scatter plot of the Cherenkov angle vs. particle momentum, a performance plot of the HMPID as recorded during pp collisions. The cluster multiplicity in these events is much lower than in Pb-Pb collisions, making pattern recognition easier. The three bands represent the measured Cherenkov angles for pions, kaons and protons, respectively, aligned with the theoretical curves. The excellent agreement is evident. Figure 7-10 shows the PID performance in two momentum intervals: $2.5 < p_T < 2.6$ GeV/c and $3.8 < p_T < 4.0$ GeV/c, in proton collisions at a center-of-mass energy of $\sqrt{s} = 13$ TeV, derived from the scatter plot in Fig. 7-9. The area under each Gaussian curve in various momentum intervals allows for the measurement of the relative abundances of pions, kaons, and protons, providing the determination of the charged hadron spectra. Thus, the HMPID fulfilled the PID role designated within the ALICE Experiment from 2010 to 2015. The momentum range in which the HMPID operates also enables the detection of deuteron in the 3−10 GeV/c.

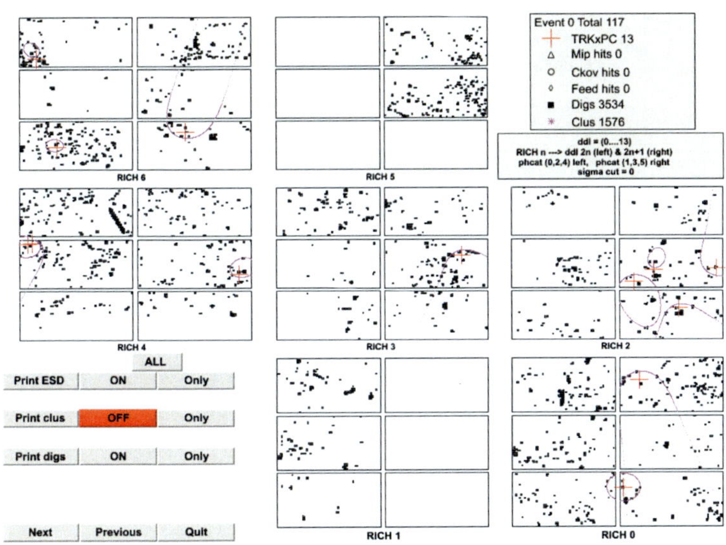

Figure 7-8. Event display of a Pb-Pb collision at LHC for the entire HMPID. Image credit: CERN.

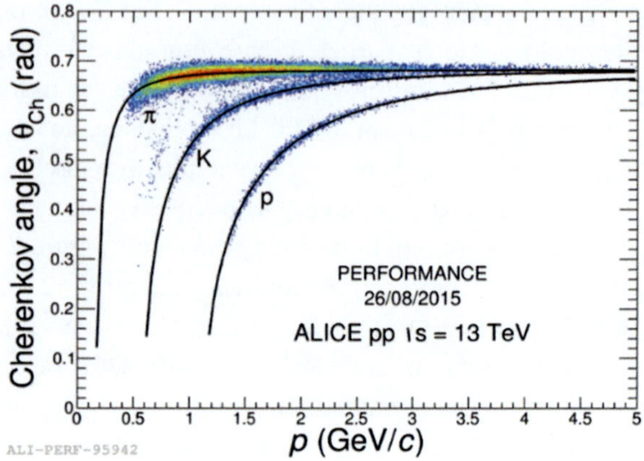

Figure 7-9. HMPID Cherenkov angle vs. track momentum for pp collisions at √s = 13 TeV data. The continuous lines represent theoretical trends. Image credit: CERN.

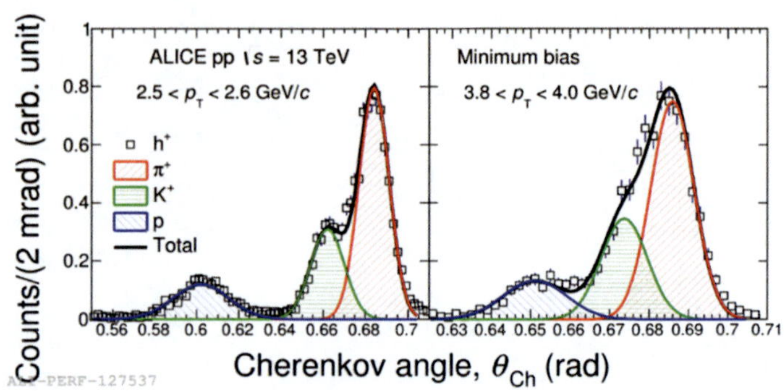

Figure 7-10. Distributions of the Cherenkov angle measured in the HMPID for positive tracks with pT in two intervals, 2.5−2.6 GeV/c (left) and 3.8−4.0 GeV/c (right), for pp collisions at √s = 13 TeV. The histograms have been scaled to have similar maximum values. Image credit: CERN.

In Fig. 7-11, the squared mass spectrum is shown. The bump below the "d" indicates all the deuterons identified in the data sample within the momentum range of the HMPID's operation [Barile, 2014. 1-8] ALICE Collaboration. https://arxiv.org/abs/1411.1941).

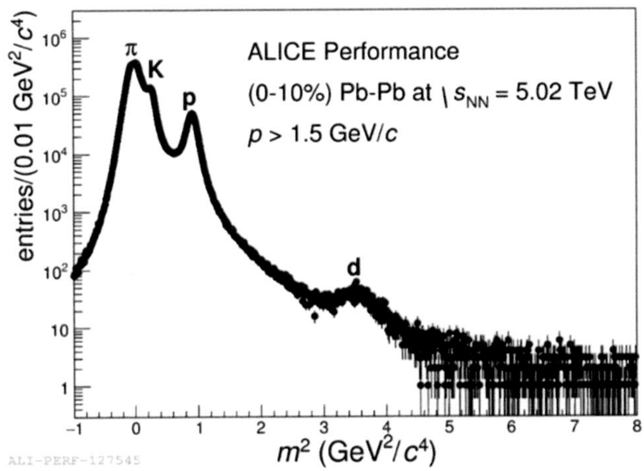

Figure 7-11. Particle squared mass distribution obtained from the Cherenkov angle measured in the HMPID, combined with the momentum information provided by the ALICE tracking devices, in central (0-10%) Pb−Pb collisions at $\sqrt{s_{NN}}$ = 5.02 TeV. Image credit: CERN.

7.6.2. The MWPC for the ALICE TPC

The basic principle of TPC operation is described in Chapter 5 (see Fig. 5-8). For the ALICE experiment, a Time Projection Chamber measuring 5.6 meters in diameter and 5.5 meters in length was selected as the primary tracking and particle identification detector. It was chosen for its ability to handle high charged-particle multiplicities, with approximately 8000 particles per unit of rapidity (dN/dy = 8000), despite known limitations related to readout speed and data volume [Dellacasa, 2000, 1-226].

A key element of this detector is the gating grid, located in front of the cathode wire plane of the endcap MWPCs (see Fig. 7-12). In the absence of a trigger, the gating grid is biased with alternating voltages to prevent electrons from the drift volume from reaching the amplification region, and to stop ions created in previous avalanche processes from moving back in the drifting volume. This is crucial because escaping ions can accumulate in the drift volume, leading to severe distortions of the drift field.

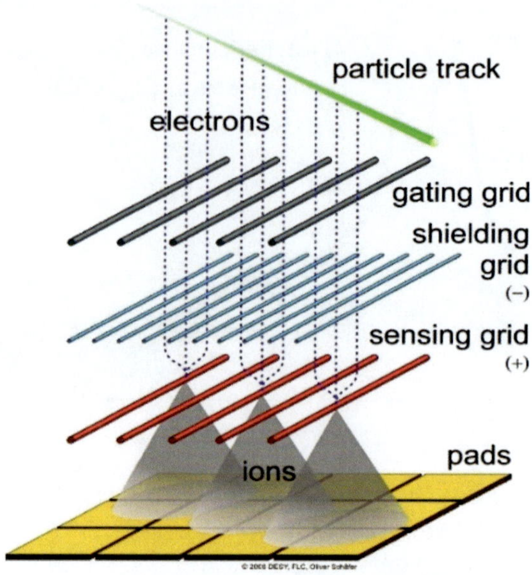

Fig 7-12. Operation principles of a Time Projection Chamber. Reproduced from https://www.lctpc.org/e8/e46/ with permission from Oliver Schäfer, DESY.

Figure 7-13 shows the TPC being lowered into its final position within the ALICE space frame.

Each of the two endcaps of the ALICE TPC is equipped with eighteen sectors, consisting of inner and outer MWPCs arranged at increasing radii from the beam pipe. These chambers are fitted with pad-segmented cathodes (Fig. 7-14), which provide the X and Y coordinates of the electron clusters generated by charged particles passing through the TPC volume. By also measuring the electron drift time, the Z coordinate can be determined, thereby locating the cluster within the sensitive volume of the TPC.

Fig. 7-15 offers a striking view of one half of the TPC's internal volume, mirrored by the central membrane to give the impression of the detector's full length. Within this volume, both the drift electric field and the magnetic field, aligned parallel to the beam pipe, are present.

Figure 7-13. The installation of TPC in the ALICE space frame in 2008. Image credit: CERN.

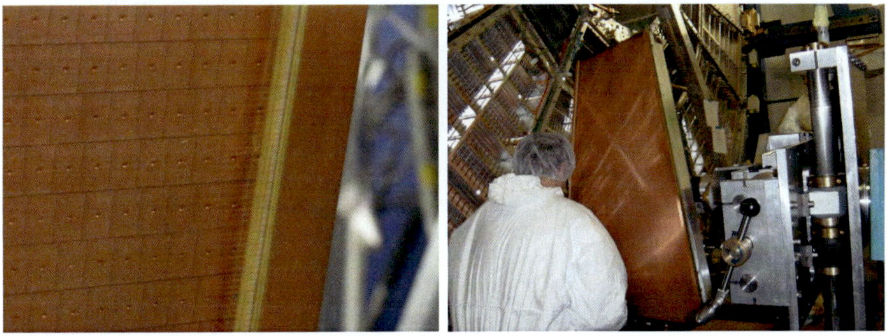

Figure 7-14. On the left panel is shown a detail of the TPC MWPC with the pad segmented cathode and anode wires. On the right an entire sector equipped with inner and outer MWPCs. CERN image.

Figure 7-15. View of one-half of the drift volume of TPC. Image credit: CERN.

The square luminous aperture at the center of the right edge of the image is the mirrored reflection of the opening through which the photograph was taken. A portion of the aperture frame on the TPC endcap is visible in the bottom left corner of the image, corresponding to one of the sectors that has not yet been equipped with MWPCs.

It is within this gas-filled volume that charged particles produce electron clusters along their paths.

As an example, Fig. 7-16 presents an impressive image of particle tracks recorded by the TPC during a single Pb–Pb collision event.

A computer-based data processing of electron clusters enables the reconstruction of particle tracks within the TPC volume. The precise measurement of the drift time is achieved using a Time-to-Digital Converter with picosecond precision. The curvature of the resulting track in the magnetic field allows the measurement of both the particle's momentum and the sign of its electric charge.

The TPC ensures efficient track finding and high tracking precision for momentum measurement, precision that is essential for the Particle

Identification (PID) task. By integrating with other detectors, ALICE achieves effective PID over a wide momentum range, from hundreds of MeV/c up to hundreds of GeV/c.

Figure 7-16. A central Pb–Pb collision recorded with the ALICE TPC. Image credit: CERN.

Figure 7-17 shows the PID performance of the TPC as a function of rigidity. The specific energy loss (dE_l/dx) due to ionization in the TPC (Y-axis) is plotted versus rigidity (p/z) on the X-axis. Both identified particles and antiparticles are shown.

The PID capabilities of the experiment are a prerequisite for studying the underlying dynamics of the strong interaction during the phase transition from the quark–gluon plasma to ordinary matter, the protons and neutrons, together with electrons, that make up the matter we are composed of.

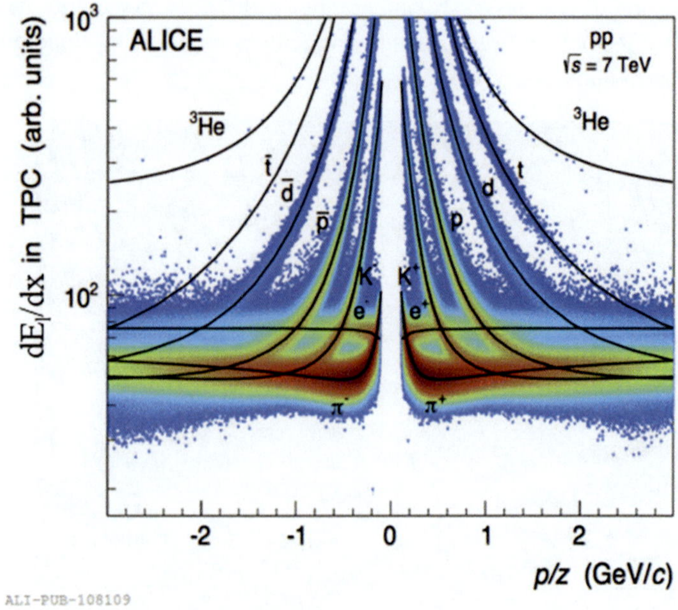

Figure 7-17. Specific energy loss (dE$_I$/*dx) vs. rigidity (momentum/charge) for TPC tracks from pp collisions at* $\sqrt{s}$ *= 7 TeV. The solid lines represent a parametrization of the Bethe-Bloch curve. Image credit: CERN.*

7.6.3. The MWPC for the ALICE Transition Radiation Detectors

The principle of TRD operation is described in Section 6.2. The effective detection and use of transition radiation for particle identification and energy measurements became technically feasible when large-area, regularly spaced, or foam radiators were coupled with MWPCs for X-ray detection.

To improve X-ray detection, a percentage of xenon is typically added to the gas mixture circulating through the MWPCs, and fast, efficient electronics are employed for signal processing.

Collider experiments offer many examples of this approach, one of which is the ALICE TRD. Its position within the detector is shown in Fig. 7-4.

The primary goal of the ALICE TRD is to provide electron identification in the central barrel at momenta above 1 GeV/c, where pion rejection via energy loss measurement in the TPC is no longer sufficient. In this momentum range, electrons emit transition radiation, whereas pions do not. Consequently, the addition of the TRD significantly broadens the physics reach of the ALICE experiment [Cortese, 2001, 1-268].

In Pb–Pb collisions, during the quark–gluon plasma (QGP) phase, electrons do not interact strongly with the medium; therefore, their production makes them excellent messengers for probing the QGP.

Figure 7-18. One (out of 18) of the TRD modules during its insertion into the ALICE space frame. CERN image.

Figure 7-18 shows the insertion of one TRD module into the ALICE space frame. The ALICE TRD consists of 18 modules, each module based on a stack of six TRD layers. Each layer is based on a MWPC with pad-segmented cathodes and a drift region between the cathode plane and the anode wire plane.

The radiator has a thickness of 4.8 cm and is composed of a combination of foam and fiber mats in a sandwich structure. To enhance the conversion of TR X-rays, a fraction of xenon is added to the MWPC gas mixture. Signal processing is performed by a 7.5 Ms/s waveform digitizer, which enables the separation of X-ray signals from the dE_l/dx contributions; these components can be clearly identified in the time spectrum of the signals from the drift chamber, as shown in Fig. 7-19.

The peak at early times originates from the amplification region, while the plateau at intermediate times corresponds to the drift region. The average pulse height is higher for electrons because, in this momentum range, the specific energy loss of electrons is greater than that of pions. The peak at the end (at 2.5 μs of drift time) represents the converted X-ray photons as they enter the drift region.

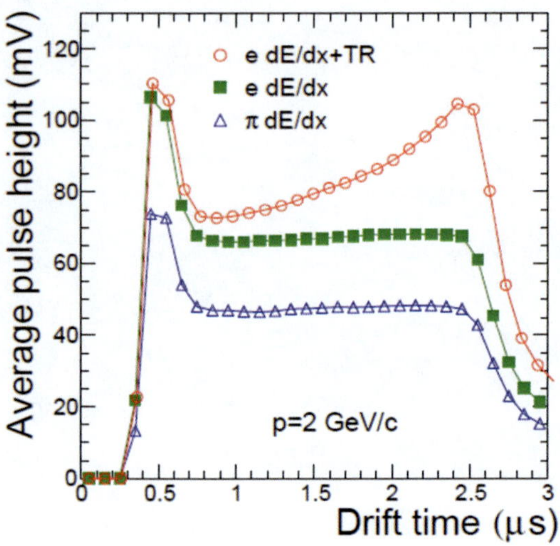

Figure 7-19 Average pulse height vs. drift time for pions and electrons (with and without radiator) with momenta of 2 GeV/c, measured with the ALICE TRD. Image credit: CERN.

A detailed description of the ALICE TRD performance can be found in: [Acharya, 2018, 88-127].

Concluding remarks

The use of MWPCs and drift chambers (DCs) in the TPC, TRD, and HMPID detectors of ALICE, together with other detection techniques, has been crucial in providing the experiment with highly effective particle identification for studies of the quark–gluon plasma. This success, however, would not have been possible without dedicated, high-performance electronics.

CHAPTER 8

PROBING THE COSMOS: APPLICATIONS OF MWPCS IN ASTROPHYSICS

Abstract

In this section, several examples of MWPC applications in experimental astrophysics are given.

Introduction

Experimental astrophysics involves observing and measuring celestial phenomena using a variety of instruments. Its goal is to investigate and understand the properties and nature of astronomical objects, such as stars, galaxies, black holes, and the universe itself, thereby contributing to the broader field of cosmology. This branch of science relies on multiple technologies, including telescopes, satellites, ground-based observatories, and particle detectors. These instruments collect data across the electromagnetic spectrum, visible light, radio waves, X-rays, and gamma rays, as well as from cosmic-ray particles, providing crucial insights into the characteristics and behaviour of celestial objects.

For example, the question of whether matter and antimatter exhibit global symmetry or asymmetry in the universe has been a longstanding problem in cosmology. It is well established that, within our Milky Way galaxy, an asymmetry exists: matter overwhelmingly dominates over antimatter. By measuring the flux and spectrum of antiparticles in space, scientists can gain valuable information to help address this fundamental question.

Astrophysicists conduct experiments underground, at ground level, or in the upper atmosphere to study the origins and propagation of cosmic radiation through space.

Charpak's Multi-Wire Proportional Chambers have been, and continue to be, employed in various experimental contexts, particularly for detecting high-energy cosmic rays, which may originate even from outside our galaxy and constantly bombard Earth's atmosphere. MWPCs have also been used in X-ray astronomy flight missions, leading to important discoveries. Below are some selected examples.

8.1. The MWPCs in the cosmic rays study

8.1.1 Ground based experiments

There are a large number of small cosmic ray stations located in laboratories and educational centers around the world. These are primarily used for detector calibration and educational purposes. In addition, several dedicated cosmic ray detectors are focused on studying the origin and physics of cosmic rays, with the largest facilities devoted to the study of extensive air showers (EAS).

EAS are initiated when ultra-high-energy cosmic rays, typically protons or atomic nuclei, collide with particles in Earth's atmosphere. This collision triggers a cascade of secondary particles, which continue to interact as they propagate, creating a shower that can span kilometres and involve billions of particles. The energy of the primary cosmic ray is distributed among this multitude of secondary particles.

Extensive air shower experiments use ground-based detectors to capture these secondary particles. These detector arrays employ various types of sensors, such as scintillators, Cherenkov detectors, and particle detectors, including MWPCs. Many of these experiments utilize gaseous detectors like MWPCs for tracking and detection. A few illustrative examples are provided below.

Several techniques are used in EAS experiments to collect data and make meaningful observations:

- Ground arrays: A network of particle detectors spread over a wide area. By measuring the timing and intensity of particle arrivals at different detectors, researchers can reconstruct the direction and energy of the primary cosmic ray.

- Cherenkov detectors: These detect Cherenkov radiation emitted when charged particles travel through a medium faster than the speed of light in that medium. This phenomenon helps determine properties of the shower particles and infer characteristics of the primary cosmic ray.
- Fluorescence detectors: These detect faint ultraviolet light emitted by nitrogen molecules excited by the charged particles in the air shower. This technique provides information on the longitudinal development of the shower and the energy of the primary cosmic ray.
- Hybrid detectors: These combine multiple techniques, such as ground arrays and fluorescence detectors, offering more comprehensive data and improving measurement precision.

A few illustrative examples of such experiments are presented below.

CLUE Experiment (1994–2001)

The Cherenkov Light Ultraviolet Experiment (CLUE) consisted of four telescopes located at 2200 meters above sea level on the island of La Palma, Canary Islands (see, for example, [Alexandreas, 1998, 488-491]). The CLUE experiment studied primary cosmic rays with energies $E > 2$ TeV by detecting ultraviolet Cherenkov light (190–230 nm) produced by atmospheric showers.

CLUE was originally designed as an array of eight ultraviolet Cherenkov telescopes. Each telescope consisted of an F/1, 1.8-meter aluminized mirror mounted on an alt-azimuth mount housed inside a modified standard 6-meter container, which could open like a clamshell during data-taking.

The detectors used in the UV telescopes were photosensitive MWPCs filled with TMAE vapours at 32°C. These solar-blind gaseous photomultipliers enabled the detection of UV radiation from extensive air showers at night, even under bright moonlight. Maintaining a high and uniform chamber temperature posed a significant challenge due to the site's frequent cold, strong winds. This was initially addressed by using multiple computer-controlled heating elements for earlier detector versions.

In the updated chamber design, the pad readout system (576 pixels) was placed outside the chamber, sealed on the readout side with a high-resistivity germanium layer to maintain the internal electric field. This design aimed to improve detector control but unexpectedly also led to increased gas gain in the chamber.

The CLUE experiment was the first to demonstrate that TMAE-based MWPCs could operate reliably under harsh environmental conditions. The images obtained with these detectors were analysed using sophisticated algorithms and statistical techniques. The results provided valuable insights for theoretical models, simulations, and the design of future experiments.

KASCADE Ground (2000–2012)

The KASCADE (KArlsruhe Shower Core and Array DEtector) ground experiment was designed to measure the composition of cosmic rays in the energy region around the knee, at energies of several 10^{15} eV [Antoni, 2001, 245-260]. This region of cosmic ray energies remains poorly understood, not due to a lack of theories, but because accurate measurements are extremely difficult at such low flux levels, typically less than 1 particle per square meter per year. At these fluxes, direct measurements from satellites or balloons are impractical.

KASCADE consisted of three main components: the array, the muon tunnel, and the central detector.

The array comprised 252 detector stations equipped with scintillation counters, covering an area of approximately 200×200 m^2. The muon tunnel, with an area of 5.4×48 m^2, was largely shielded with absorber material on top and outfitted with layers of streamer tubes.

The central detector was the most complex instrument. Its primary component was a 16×20 m^2 hadron calorimeter, utilizing 4,000 tons of iron as the absorber material. The calorimeter consisted of eight layers, each fully instrumented with room-temperature liquid ionization chambers filled with tetramethylsilane (TMS).

Beneath the calorimeter were two layers of MWPCs (multi-wire proportional chambers), each with a sensitive area of 122 m^2. These chambers functioned as position- and direction-sensitive muon counters, designed to detect muons with an energy threshold above 2 GeV. By employing the MWPCs, it was possible to determine the number of reconstructed muons and gather detailed data on their hit patterns.

Thanks to its diverse array of detectors, KASCADE excelled at reconstructing numerous shower observables. These included the shower direction, core position, electron number, and truncated muon number (integrated over a core distance range of 40–200 meters), among others.

In its later phase, the KASCADE project successfully evolved into KASCADE-Grande [see, for example, [Apel, 2010, 202-216].

8.1.2 Balloon-borne Experiments

High-altitude balloons carrying cosmic ray detectors can reach altitudes of up to 30 km, with typical flight durations of around 20 hours. These experiments have provided critical early data on cosmic ray fluxes at different altitudes and latitudes.

In this section, we highlight several balloon flight experiments that employed MWPCs alongside other instruments to investigate this fundamental phenomenon.

A common feature of all these experiments was the use of a rigidity spectrometer, typically incorporating a superconducting magnet. The spectrometer was combined with specific particle detectors tailored to the scientific objectives of each successive high-altitude balloon flight.

It is worth noting that terminology in older astrophysics literature can sometimes be confusing: in early papers, “MWPC” was occasionally used to refer to an array of single-wire counters. Here, we will focus exclusively on experiments that utilized classical Charpak-type MWPCs.

BESS (1993–1995)

The Balloon-borne Experiment with a Superconducting Spectrometer (BESS) was designed to study the matter–antimatter asymmetry in the early Universe by precisely measuring antiprotons, positrons, and gamma rays, and by searching for primordial antimatter in cosmic rays. See, for example, [Ajima, 2000,71-100].

The BESS spectrometer consisted of four principal components: a thin superconducting solenoid (transparent to light, low-Z cosmic particles), drift chambers for particle tracking, time-of-flight scintillation counters, and a data acquisition system.

The particle tracking system was based on multiwire chamber technology. It included a central drift chamber for precise measurement of cosmic ray trajectories in the magnetic field, along with inner and outer drift chambers for fast pattern recognition based on angular deflection. A key technological feature was that the entire system operated at 1 atm in a sealed mode, without gas circulation.

The central cylindrical drift chamber determined particle trajectories in the plane transverse to the magnetic field, enabling continuous tracking with a precision of 200 μm in the uniform magnetic field.

The inner and outer drift chambers provided rapid identification of the charge sign and angular deflection, serving primarily for onboard event selection. These chambers had radially double-layered configurations, achieving an azimuthal resolution of 300 μm and an axial (z) position resolution of 230 μm. Thanks to this design, BESS had a unique capability to measure the direction of point-source gamma rays: the solenoid wall acted as a converter, and the resulting electron–positron pairs were tracked in the magnetic field with an angular resolution better than one degree in the energy range of 0.1–100 GeV.

During its 1993 flight, BESS achieved the first detection of mass-identified low-energy antiprotons below 500 MeV. In subsequent flights in 1994 and 1995, it measured an improved antiproton flux over a wider energy range with higher statistical precision [Yoshimura, 1995, 3792-3796], [Moiseev, 1997, 479-490].

TS93 (1993)

The balloon-borne experiment TS93 was optimized for the detection of antimatter in primary cosmic rays. It was based on a superconducting magnet spectrometer designed to identify and separate electrons from antiprotons, and positrons from protons (which share the same electric charge sign but have very different masses) in the energy range of 5–60 GeV [Barbarito, 1995, 588–600].

To enhance particle identification capability, a Transition Radiation Detector using MWPCs was integrated at the top of the spectrometer. The payload was launched from Fort Sumner, New Mexico, USA, by NASA, reaching an average altitude of about 36 km, corresponding to only 5 g/cm^2 of residual atmosphere, for a total flight time of 25 hours.

The TRD consisted of a stack of 10 modules. Each module consisted of a MWPC filled with a xenon-based gas mixture, along with radiators made of 25 μm thick polyethylene bags filled with mats of carbon fibers. On average, for a radiating particle crossing the 10 layers, about 10 X-ray photons were detected, while for non-radiating particles only about two photons were detected. Signal processing for each wire was based on fast cluster-counting electronics with a simple 2-bit counter, sufficient since no more than two clusters per wire were expected. This low-power, simple, and effective signal processing method was also well-suited for satellite missions.

The measured positron flux was higher than predicted by models considering only secondary production. Today, this excess has been confirmed by other experiments, e.g. HEAT, Fermi-LAT (https://ecap.nat.fau.de/indexphp/research/gamma-ray-astronomy/fermi-lat/), HESS, CAPRICE94, PAMELA (electrons only), and AMS-02 (https://www.home.cern/science/experiments/ams). Among various explanations, the decay or annihilation of dark matter is also invoked.

More details on the presence of primordial dark matter or antimatter in cosmic rays are provided in the Chapter 10.

CAPRICE (1994–1998)

The primary objective of the Cosmic Antiparticle Ring Imaging Cherenkov Experiment (CAPRICE) was to achieve precise measurements of antimatter, specifically antiprotons and positrons (see, for example, [Boezio, 2001, 669-674]).

It was a balloon-borne experiment, using the same spectrometer as in TS93 but equipped with different detectors for particle identification. The experiment aimed to conduct a thorough search for antinuclei as well as light isotopes in cosmic radiation. A significant portion of the antiprotons and positrons reaching Earth originates from high-energy interactions between cosmic ray nuclei and the interstellar medium. Analyzing their spectra provides valuable insights into the origins, production mechanisms, and propagation patterns of cosmic rays across our galaxy.

The detection of any flux exceeding the predictions of standard cosmic ray propagation models, such as the Leaky Box Model, could indicate the presence of exotic sources of antimatter. Since the predictions of propagation models diverge significantly above 10 GeV, more refined measurements were needed in this energy range to improve our understanding.

The CAPRICE setup consisted of several key components: a superconducting magnet spectrometer, a time-of-flight detector, a silicon-tungsten electromagnetic calorimeter, and two types of multiwire chambers, one for tracking and one as a RICH detector.

The tracking system included three sets of drift chambers, each with 10 layers of hexagonal cells, achieving a spatial resolution of approximately 100 μm. A distinctive feature of the RICH detector was a MWPC filled with TMAE. This detector was operated in two configurations: proximity-focused geometry during the 1994 flight, and with an added mirror in the 1998 flight. In the 1994 configuration, a solid NaF radiator was used, producing about 10 photoelectrons per incident ultra-relativistic particle. This marked the first-ever use of a RICH detector in a space experiment capable of detecting unit-charge particles like antiprotons.

In 1998, NaF was replaced with a gaseous radiator (C_4F_{10}), and a mirror was integrated into the RICH system. In both cases, the RICH detector was equipped with an automated temperature control system for the gas circuit, the MWPC frame, and the radiator itself, designed to prevent condensation outside the TMAE container. Notably, the entire RICH detector was thermally insulated from the rest of the payload.

CAPRICE's advanced tracking system, comprising drift chambers and the innovative RICH detector, enabled precise particle identification and trajectory reconstruction. This allowed researchers to distinguish between different types of particles and antiparticles, leading to the detection of antimatter components in cosmic rays. The experiment's findings not only confirmed the presence of antimatter particles in cosmic rays but also provided insights into their abundance and distribution in the upper atmosphere. The ability to identify and characterize these particles at various altitudes contributed to refining models of cosmic ray propagation and interaction processes.

HEAT experiment (1995)

Like other similar flight missions, the High-Energy Antimatter Telescope (HEAT) was developed to measure the abundances and energy spectra of positrons and antiprotons, as well as the isotopic composition of cosmic ray nuclei (see, for example, [Barwick, 1997, 34-52]).

The combination of the spectrometer with time-of-flight scintillators, transition radiation detectors, and an electromagnetic shower counter provided highly effective discrimination against the large background of protons. MWPCs with cathode strip readout were used in the TRD, which consisted of six modules with polyethylene fibre radiators.

To distinguish the X-ray signal from ionization tracks, a similar approach was employed as described in Chapter 7.4.3: transition radiation X-ray photons generated well-localized ionization clusters inside the gas, producing signals with different amplitudes and timing compared to those from the ionization track. The tracking capability of the TRD was also used to identify and reject delta rays. For more refined analysis, a maximum likelihood method was applied (see [Barwick, 1997, L191-L194]). The

HEAT balloon mission successfully collected data in 1994 and 1995, performing measurements of cosmic-ray electrons and positrons below 100 GeV.

The HEAT measurements reinforced earlier findings, showing that the positron fraction does not increase in the 10–50 GeV energy range, which remained consistent with theoretical predictions (see [Protheroe, 1982, 391-397]).

Note for Curious Readers: *Improved Precision in Cosmic Ray Antimatter Measurements: Insights from AMS and PAMELA*

Satellites equipped with particle detectors are capable of measuring the energy and composition of cosmic rays over multi-year periods. This capability has significantly improved the precision of antimatter measurements in primary cosmic rays.

To date, the Alpha Magnetic Spectrometer (AMS) and the PAMELA experiment, thanks to their high sensitivity, have not identified any candidates for primordial light nuclei. Further information on this topic can be found in Section 10.3.

8.1.3. Satellite experiments

Two examples of satellite experiments that employed MWPCs are CRN, which investigated the composition of cosmic ray nuclei, and GRANAT, which focused on detecting X-ray sources.

CRN Experiment (1985)

Until 1985, balloon missions had provided a substantial amount of information on the composition of galactic cosmic rays up to energies of several tens of GeV/amu (where 1 amu = 1.66×10^{-27} kg). However, at higher energies, data became increasingly scarce, leading to limited knowledge in the TeV/amu range and beyond. Due to the rapidly decreasing flux of cosmic rays with increasing energy, studies at these energies require large-area detectors and long exposure times.

These requirements were met for the first time during the Shuttle Challenger mission in 1985. The instrument installed, called the Cosmic Ray Nuclei detector (CRN) [Grunsfeld, 1988, 381-390], was a telescope of unprecedented size compared to standard balloon missions: 2.7 meters in diameter, 3.7 meters in length, and weighing 2400 kg. It consisted of two Cherenkov detectors, with a TRD positioned between them, all surrounded by scintillation detectors. The TRD was a multilayer structure of radiators and MWPCs, which also served as a hodoscope for determining particle trajectories.

During the flight, the spectral shapes of elements such as carbon (C), oxygen (O), neon (Ne), and iron (Fe) were measured. A key finding was that the spectra of nitrogen (N), magnesium (Mg), and silicon (Si) did not follow the trend predicted by existing models of cosmic-ray acceleration.

GRANAT flight experiment (1989-1997)

Another remarkable example of the use of MWPCs in space research was the GRANAT flight mission (see https://heasarc.gsfc.nasa.gov/docs/granat/granat_about.html, and https://heasarc.gsfc.nasa.gov/docs/granat/granat.html).

The GRANAT project was developed in the 1980s through a collaboration between the USSR, France, Denmark, and Bulgaria. This specialized astrophysical observatory carried seven instruments: the French gamma-ray telescope with a coded aperture (SIGMA), the X-ray telescope with a coded aperture (ART-P), the X-ray spectrometer combined with a collimator (ART-S), the Danish X-ray all-sky monitor (WATCH), two gamma-ray burst detectors (KONUS and the French PHEBUS), and a set of instruments mounted on the Sunflower rotary platform developed jointly with Bulgarian specialists.

MWPCs were used in both ART-P and ART-S. Each instrument consisted of four modules containing specially designed, mechanically robust MWPCs sensitive in the 4–100 keV energy range.

During its five years of active operation, GRANAT collected highly valuable data. It revealed that the dominant sources of X-ray emission in the Galaxy are compact sources clustered near the Galactic Center within a few degrees of angular distance. The maps produced by the GRANAT telescopes were the most detailed X-ray maps of the Galaxy at that time. Additionally, GRANAT obtained a map of diffuse X-ray radiation.

The ART-P telescope detected quasi-periodic oscillations in the X-ray flux from X-ray novae in the constellations Musca and Perseus, as well as from the source GX 339-4, all of which are candidate black holes. These observations demonstrated that QPOs are not unique to neutron stars, but can also occur in accretion disks around black holes due to hydrodynamic and thermal instabilities (https://adsabs.harvard.edu/full/1993foap.conf..175C and https://scholar.google.com/scholar?hl=de&as_sdt=0%2C5&q=results+from+of+GRANAT+flight+experiment+&btnG=)

Over its eight years of operation, GRANAT discovered about twenty previously unknown X-ray sources, candidate black holes and neutron stars. These sources were named with the prefix "GRS" (GRANAT Source), denoting their discovery by GRANAT.

Concluding remarks

Ground-based and space-based cosmic ray experiments have made significant contributions to our understanding of astrophysical processes. MWPCs have played a crucial role in many of these experiments, providing precise detection and measurement capabilities.

By studying the properties of cosmic rays and X-rays, scientists gain valuable insights into extreme environments and processes occurring throughout the universe, including the mechanisms of particle acceleration in supernova remnants, the structure and dynamics of cosmic magnetic fields, and the potential signatures of dark matter.

As technology continues to advance, these experiments are poised to uncover even more of the universe's mysteries, deepening our understanding of the fundamental building blocks of nature.

CHAPTER 9

OPTICAL CHAMBERS: POTENTIAL APPLICATIONS AND FUTURE DIRECTIONS

The earliest applications of optical chambers (see paragraph 5.4) were relatively limited, focusing on areas such as imaging cosmic rays and beam particles, autoradiography [Dominik, 1989, 779-787], low-energy nuclear decay [Miernik, 2007,194-197], and dosimetry [Titt ,1998, 85-99]. In these cases, the low acquisition rate of CCDs did not pose a significant limitation.

Notably, the first optical chambers provided only two-dimensional (xy) images of particle tracks, which is why they were referred to as "projection chambers."

Later, a PMT was integrated into the setup, positioned near the CCD chamber. This addition provided signals with high time resolution, enabling a rough estimation of the position of the shower in the Z direction. However, interest in optical readout techniques waned for some time.

A renewed momentum in this field emerged with the invention of advanced amplification structures, such as MICROMEGAS [Charpak, 1995, 1-30], GEM [Sauli, 1997, 531-534], and Thick GEM (THGEM) [Periale,2002, 377-383]), which unlocked new possibilities for imaging techniques.

By fortunate coincidence, it was also discovered around this time that mixtures of noble gases with CF_4 emit light in two distinct spectral intervals: 260–320 nm (similar to TEA) and 520–720 nm (the visible region, similar to TMAE) (see Fig. 9-1). This discovery opened new opportunities for their use in optical cameras.

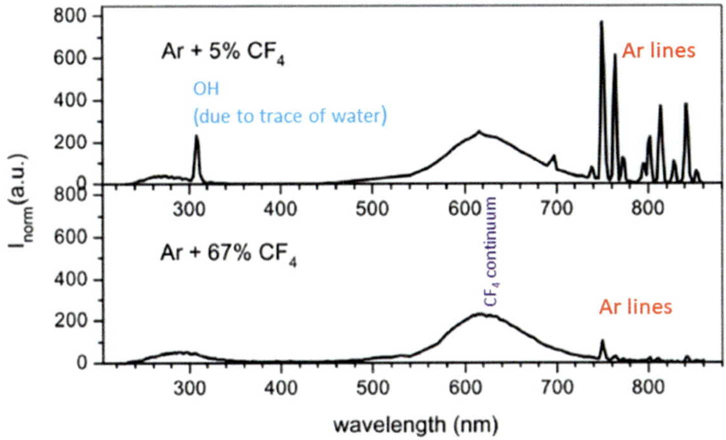

Figure 9-1. Avalanche emission spectra measured in mixtures of Ar with 5% and 67% of CF_4. Adapted from [Fraga, 2003, 88-92].

Although there is evidence that the light yield of these mixtures is slightly lower compared to TEA (see Fig. 5-10), they remain attractive primarily due to their emission in the visible spectrum, allowing the use of conventional optical systems. However, it is important to note that CF_4 has several disadvantages compared to TEA. It is an electronegative gas, and some of its dissociation products produced during avalanches are chemically reactive.

Around the same time, several groups developed prototypes of optical TPCs filled with noble gas and CF_4-based mixtures. Additionally, other gas mixtures were explored, such as $Ar+CO_2+N_2$ (in high-pressure TPCs) and pure noble gases (see NEXT). The choice of gas mixtures depends on the specific experimental requirements and the operational characteristics of the charge/light amplification structures used, such as MICROMEGAS, GEM, THGEM, PPAC, or MWPC.

There are two main directions for optical chamber applications: as 2D imaging detectors or as 3D imaging devices (e.g., TPCs).

2D imaging systems with CCD or CMOS (Complementary Metal Oxide Semiconductor) readouts are particularly useful for specialized applications, such as neutron imaging

Until recently the readout speed of existing optical imaging systems (such as CCD, scientific CMOS, etc) was insufficient for determining the third coordinate with the necessary spatial accuracy. To overcome this problem, a few groups developed TPC prototypes with combined optical and charge readout [Brubauer, 2018, 913-918[, [Deisting, 2021, 1-39] or used detailed time information from the PMTs (or SiPM) installed additionally to the optical system.

The advent of new, ultra-fast image-taking devices, such as Timepix, is expected to significantly transform the field. Future TPCs will be capable of reconstructing ionizing events in 3D with high spatial resolution, using only one or a few optical cameras.

This raises an important question: why are optical chambers becoming so popular when 2D and 3D sensitive gaseous detectors with electronic readout already offer excellent imaging capabilities? The growing popularity of optical chambers can be attributed to two main reasons:

1. External Placement of Optical Cameras: Optical cameras are positioned outside the sensitive volume of the TPC. This allows the use of pure gas compositions and minimizes the number of electrical feedthroughs required, reducing complexity and potential sources of contamination.
2. High Granularity and Sensitivity: Optical sensors provide high granularity, exceptional sensitivity, and an excellent signal-to-noise ratio. With appropriate lenses, they can also capture images over large surface areas, further enhancing their versatility.

Below are a few examples of ongoing promising development in optical chambers.

9.1. Optical chamber for neutron imaging

Neutron radiography is a technique used to test reactor materials and perform non-destructive examinations of nuclear fuel or radioactive waste in environments with very high γ-flux. In the work of Cools [Cools, 2023, 1-6] an optical system for neutron imaging was developed (see Fig. 9-2). This device is designed to enable real-time neutron radiography, achieving a spatial resolution of approximately 0.1 mm while exhibiting exceptional gamma rejection capabilities.

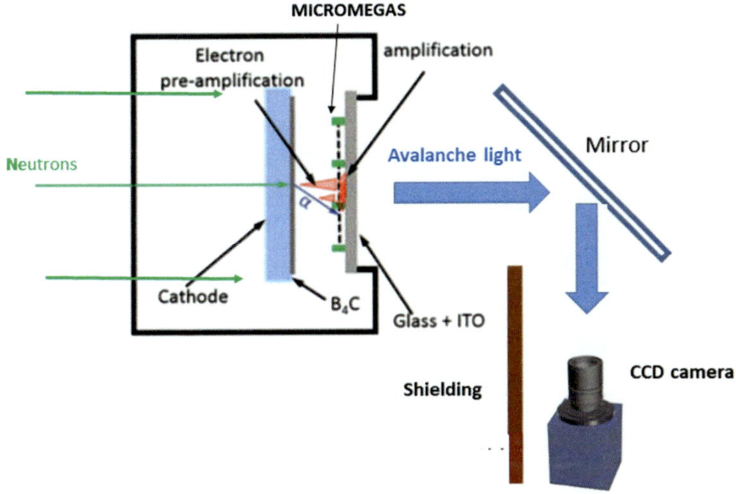

Figure 9-2. Schematic of the 2D Optical Chamber for Neutron Imaging. Adopted from [Cools, 2023, 1-6].

The system comprises three main components: a gas chamber, a mirror, and an optical sensor, specifically, the Hamamatsu ORCA-Quest qCMOS camera. Inside the gas chamber, a MICROMEGAS detector is combined with a neutron converter made of 10B4C. The 10B4C layer facilitates the conversion of low-energy neutrons into α particles or 7Li ions, which subsequently ionize the gas.

The primary electrons generated in this process drift towards the MICROMEGAS multiplication region, where they initiate avalanches. The light emitted during these avalanches is reflected by the mirror and directed to the optical camera. This configuration protects the sensor from the high particle flux.

A comparison of the works by [Charpak, 1988, 483-486,] and [Cools, 2023, 1-6] reveals that, despite employing different light-emitting structures (MWPC in Charpak's work and MICROMEGAS in Cools' system), their underlying concepts are similar: both are 2D imaging devices.

9.2. Optical TPC

NEXT TPC

NEXT (Neutrino Experiment with a Xenon TPC) is a novel experiment aimed at searching for neutrinoless double beta decay using a radiopure, high-pressure xenon TPC enriched in ^{136}Xe (see, for example [Simon, A., 2020, 1-5] and references therein). The operational principle is as follows (see Fig. 9-3): during the double beta decay process, two electrons are emitted, which interact with the xenon gas through ionization and excitation. The primary scintillation light is detected by an array of PMTs located behind the cathode, providing the initial time t_0.

The electrons produced in the ionization process are drifted by an electric field toward the anode, where they encounter a parallel plate mesh structure with a more intense electric field. Here, each primary electron produces approximately a thousand secondary (electroluminescence) photons, which are detected again by the PMTs at a later time t_1. The time difference t_0-t_1 determines the z-position of the electron track.

Forward-moving photons are detected by an array of silicon photomultipliers (SiPMs) located behind the anode. The signals from the SiPMs are used for track reconstruction. Simulations predict that these decay events produce a distinctive feature: an ionization track approximately 30 cm long, with larger energy depositions at both ends. This unique topology is utilized for event selection and background rejection.

In the subsequent prototype, the SiPMs were replaced by electron-multiplying charge-coupled devices (EMCCDs, see Fig. 9-4).

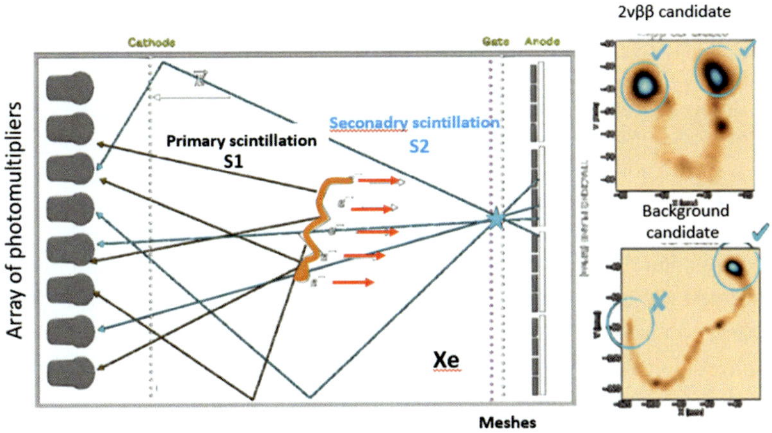

Figure 9-3. Sketch of the NEXT Time Projection Chamber, illustrating its operational principle. The upper inset on the right depicts the expected track from neutrinoless double-beta decay, while the lower inset shows a typical background track. (Adapted from https://next-experiment.org/experiment/detector/).

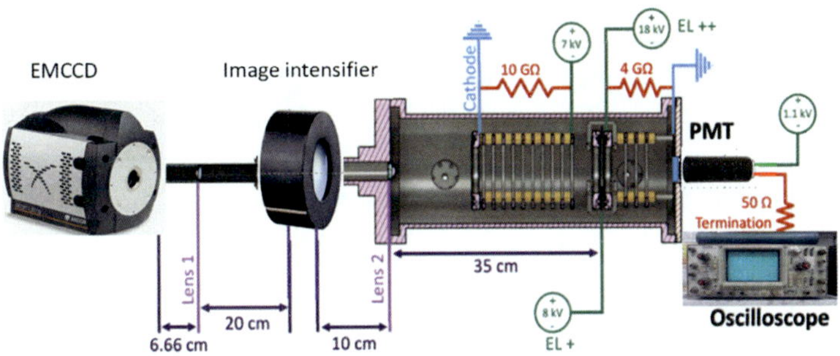

Figure 9-4. An artistic representation of one of the NEXT prototypes equipped with an ENCC (electron-multiplying charge-coupled device). Adapted from [Byrnes, 2023, 1–30].

While these cameras are not sufficiently fast to enable full 3D readout, they can capture two-dimensional projections of the track, offering a partial temporal solution for now. Full 3D imaging will be achieved in the next prototype, which will be equipped with a Timepix camera [Byrnes, . 2023, 1–30].

Cygno TPC

Directional dark matter searches aim to develop experimental techniques to determine the direction of incoming dark matter particles. This directional information provides a unique opportunity for signal identification based on topology and directionality, which is invaluable for rejecting background signals arising from internal and external radioactivity.

The CYGNO project seeks to explore the potential of an optical TPC for this purpose (see, for example, [Amaro, 2023, 1–4] and references therein). Recent advancements in CMOS technology have made this approach feasible, enabling the development of optical sensors with high sensitivity, fine granularity, and exceptionally low noise levels. These high-resolution tracking capabilities are expected to allow for precise reconstruction of particle track directions.

The CYGNO TPC employs two types of optical detectors: PMTs and Hamamatsu sCMOS cameras (Fig. 9-5). PMTs will provide fast signals, offering insights into the impinging radiation, such as the track length along the longitudinal direction and the deposited energy. Meanwhile, the sCMOS cameras will capture the 2D projection of the track. By combining data from both detectors, it will be possible to reconstruct a 3D image of the particle track.

The CYGNO collaboration plans to develop several prototypes of increasing volume before proposing a full-scale TPC.

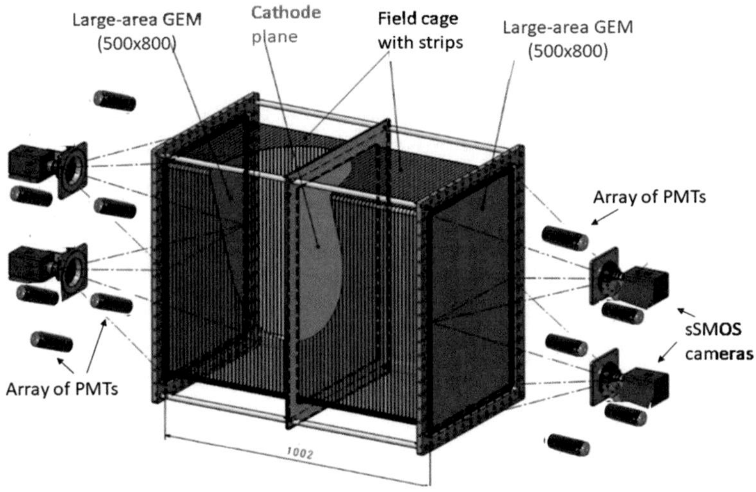

Figure 9-5. A rough sketch of the CYGNO04 demonstrator, illustrating the field cage, readout GEMs, PMs, and optical cameras (adapted from [Amaro, 2023, 1–4]).

9.3. Prototypes of High-Pressure TPC for the DUNE Experiment

The DUNE Near Detector (see, for example, ndico.cern.ch/event/1216905/ contributions/5451825/attachments/2700861/4687914/NuFact_NearDetector_2023.pdf) will consist of multiple components, each contributing complementary constraints on flux and neutrino interaction systematic uncertainties for oscillation analyses. One of these subdetectors is a high-pressure TPC, which will provide fine-grained tracking in a low-density detector, using the same target nucleus as the DUNE Far Detector.

Several options for this TPC are under consideration, such as an optical TPC similar to NEXT (but filled with an Ar-based mixture) or a TPC that combines an optical camera with charge readout of the amplification structure (see, for example, [Diesting, 2021, 1-39]). The latter approach is being pursued by several groups. The working principle of the TPC with hybrid optical and charge readout is demonstrated in Fig. 9-6. When neutrino (ν) scatters on a gas atom's nucleus two charged particles are ejected creating ionization tracks in the TPC gas. Primary electrons from the ionization track drift towards the amplification structure and produce avalanches there. An optical camera images the photons emitted during gas amplification, capturing a 2D projection of the track on the readout plane. Supplementary charge readout with coarse segmentation provides the coordinates in the third dimension.

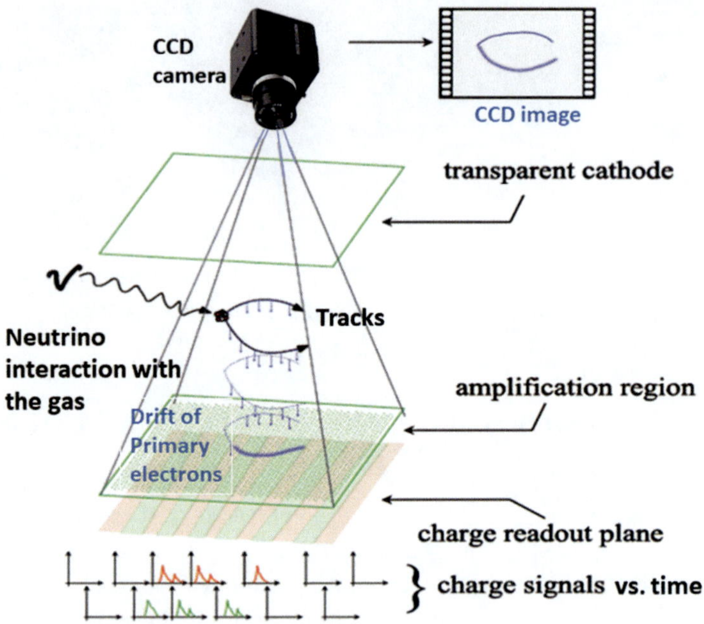

Figure 9-6. Schematic illustration of the operational principle of a TPC with hybrid optical and charge readout. Adapted from [Diesting, 2022,1-4].

The final design of the NEAR gaseous TPC will be selected based on the results of worldwide R&D efforts launched by the DUNE collaboration.

9.4. ARIADNE project

In addition to gaseous TPCs, optical readout also shows promise for noble liquid TPCs. It is expected that future LArTPCs (liquid Ar TPC) for neutrino detection and studies will reach kiloton scales. For instance, the LArTPCs proposed for the DUNE project are designed to include four modules, each with a mass of 17 kilotons.

Given this trend, it becomes crucial to develop a TPC readout system that is both cost-effective and manageable in terms of complexity. The ARIADNE (ARgon ImAging DetectioN chambEr) project aims to explore the feasibility of optical readout for large-scale LArTPCs [Lowe, 2020,1-19]. A proof of concept has been demonstrated with a prototype, as shown in Fig. 9-7.

This TPC features two inner regions: liquid argon (LAr) and gaseous argon. It operates as follows:

An ionizing event in the LAr produces primary scintillation light (S1), which is detected by an array of embedded PMTs. A high voltage applied to the anode electrode at the top of the liquid argon extracts electrons into the gaseous argon region, where a THGEM is located.

In the THGEM, the primary electrons undergo avalanche multiplication, simultaneously generating secondary scintillation light (S2). The emitted photons are recorded by both an EMCCD (electron-multiplying charge-coupled device) and the PMTs located at the bottom of the LAr TPC.

The pulses from the PMTs, the electronic signals from the segmented THGEM, and the 2D images from the EMCCD enable the reconstruction of 3D particle tracks. They also provide calorimetric information about the energy of the particles. A later baseline prototype retained the same internal structure but replaced the EMCCD with a Timepix camera (see, for example, https://medipix.web.cern.ch/timepix-and-timepix3-optical-cameras). Each pixel in the camera that detects light above a certain threshold records the x and y coordinates, as well as the time of arrival (ToA) of the primary charge.

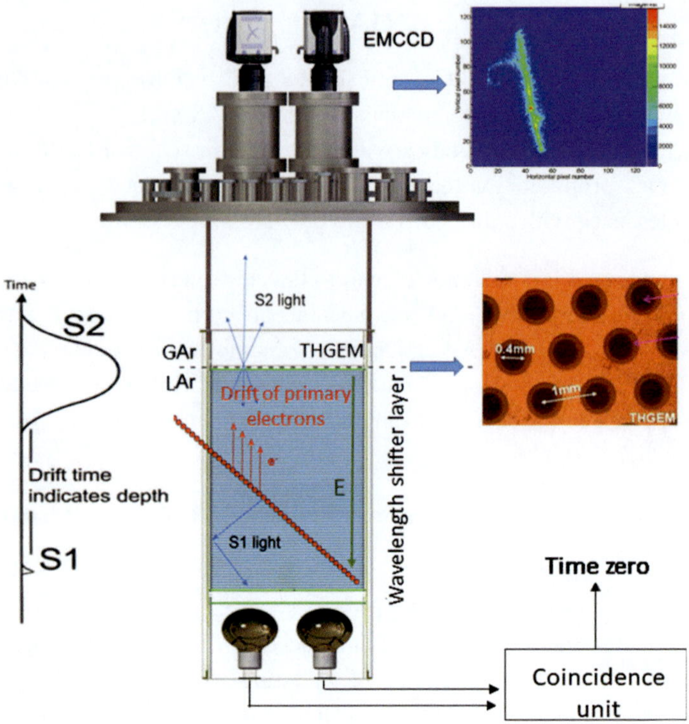

Figure 9-7. Drawing of a LAr TPC with optical readout (adapted from https://andor.oxinst.com/learning/view/article/detection-of-sub-atomic-particles-in-a-lar-tpc-detector-with-the-aid-of-an-emccd-camera).

As an example, Fig. 9-8 shows a 1-second slice of the TPX3Cam's (https://quarkphotonics.au/product/tpx3cam-single-photon-counting-camera/) continuous output data stream. Numerous tracks produced by cosmic muons during this interval are clearly visible.

The Timepix camera (or future fast cameras) enables obtaining 3D images of the track in the TPC without relying on charge information from the TPC's charge and light amplifying structure (e.g., THGEM). This innovation significantly simplifies the readout system and reduces costs by eliminating the need for multiple electrical connections and feedthroughs.

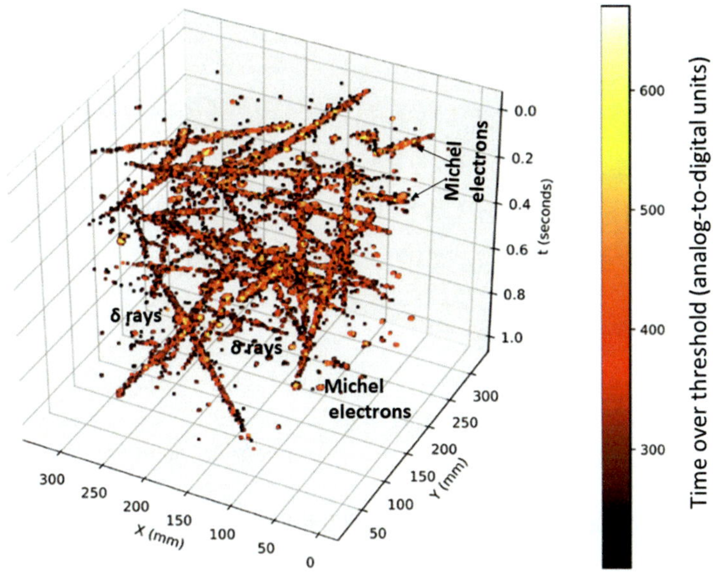

Figure 9-8. 3D visualization of muon tracks captured using an optical LAr TPC equipped with a Timepix camera (adapted from [Lowe, 2020,1-19]).

Concluding Remarks

The field of optical TPC readouts holds great promise for enhancing the precision and scalability of particle detectors. Continued advancements in photodetector technology, data processing, and material science are expected to address current challenges. Optical readouts, whether used independently or as part of hybrid systems, are likely to play a pivotal role in the next generation of experiments in fundamental physics and beyond.

CHAPTER 10

PHYSICS DISCOVERIES AND CHARPAK'S INVENTIONS: THE INTERPLAY BETWEEN PARTICLE DETECTORS AND THEORY

Why do I think Physics is the queen of the Sciences?
Because "Measurement" is its king and "Mathematics" is the wedding ring that holds them together! (GdC)

In the following paragraphs, we will highlight several significant physics discoveries made possible, in part, using Charpak chambers. Shortly after their invention, MWPCs and drift chambers were combined with magnetic dipoles in spectrometer setups to detect and identify particles in experiments at fixed-target accelerators, collider facilities, and in balloon- and satellite-borne missions investigating antimatter in cosmic rays. Notable achievements include the discovery of a particle containing the charm quark, the observation of the W and Z^0 bosons mediating the weak interaction, and contributions to the detection of antimatter in cosmic rays.

This chapter explores the intricate interplay between detector technologies and theoretical advancements, with a particular focus on Charpak's inventions.

The discussion assumes familiarity with advanced concepts in particle physics and is intended for readers with a strong interest in the subject.

10.1. Discovery of the J/ψ particle containing the new charm quark

In the 1970s, the existence of a fourth quark, alongside the up, down, and strange quarks, was theoretically proposed to explain certain intriguing experimental observations. The process involving this fourth quark became known as the GIM (Glashow–Iliopoulos–Maiani,) mechanism [Glashow, 1970, 1285-1292].

In 1974, the J/ψ particle was discovered almost simultaneously at the AGS (Alternating Gradient Synchrotron) proton accelerator in Brookhaven [Aubert, 1974, 1404-1406] and at the SPEAR e^+e^- collider at SLAC [Augustin, 1974, 1406-1408].

This concurrent observation by two separate experiments led to the particle receiving two names: J and ψ.

The AGS experiment investigated the reaction $p + Be \rightarrow J + X \rightarrow e^+ + e^- + X$, searching for new particles (J) that decay into an e^+e^- pair by precisely measuring the invariant mass of the pair with a high-resolution spectrometer. In this reaction, X denotes additional particles produced in the interaction.

The remarkably narrow mass distribution of the J/ψ around 3100 MeV, together with subsequent precise studies of its decay properties, its excited states, and other particles containing the charm quark, confirmed the unique characteristics predicted by theory. The decay behaviour, in particular, matched the expectations of the GIM mechanism, which had posited the existence of a fourth quark, named "charm." Thus, within the quark model, the J/ψ meson provided compelling evidence for a bound $c\bar{c}$ (charm-anticharm) state [Bettini, 2016, 147–153].

A review of the experimental setup (Fig. 10-1) highlights the crucial role of the tracking system, which incorporated Charpak's innovation, in enabling the detection of this new particle. On average, the spectrometer had to identify a single electron-positron pair out of every 10^8 hadron events, an extraordinary experimental challenge.

In the experimental setup, the selection of e+ + e− events was achieved by logically combining signals from the Cherenkov counters with those from the electromagnetic calorimeter, positioned behind the largest chamber (1 m × 1 m). The tracking relied on four MWPC blocks, A0, A, B, and C, on each spectrometer arm, used to identify electrons and positrons.

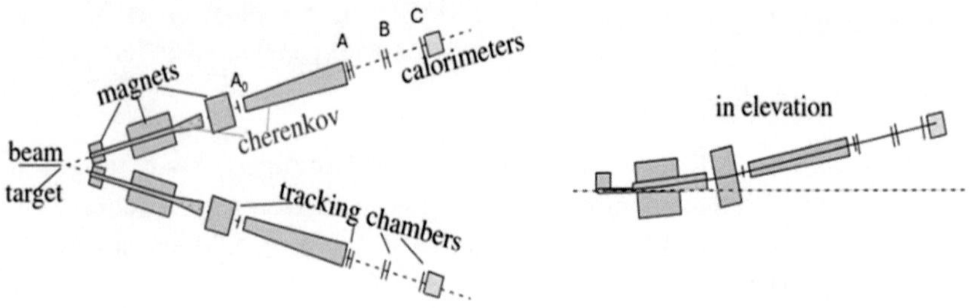

Figure 10-1. Simplified top view (left) of the Brookhaven double-arm spectrometer, and side view (right) showing the spectrometer elevated relative to the dashed horizontal line. Adapted from [Aubert, 1974, 1404-1406] and [Bettini, 2016, 147–153].

The setup included a total of eleven MWPC planes: 2×A0, 3×A, 3×B, and 3×C, each plane rotated by about 20° relative to the others to reduce multitrack confusion.

Achieving the required rejection power of 10^{-8} necessitated highly precise tracking information. It was the integration of these various detection techniques that made this groundbreaking discovery possible, with the MWPC playing a crucial role.

10.2. Discovery of W and Z^0

The W and Z^0 are the massive vector bosons mediating the weak interaction, the force responsible for radioactive decays and solar fusion reactions.

Their experimental detection marked a major milestone in validating the theoretical framework that unifies the electromagnetic and weak forces: the electroweak interaction. The existence of these vector bosons had been predicted by the electroweak theory, developed primarily by S. Glashow (1961), Salam & Ward, (1964), and S. Weinberg (1967).

A key signature of the unification of electromagnetic and weak interactions was the observation of so-called neutral current processes. These were first experimentally detected in 1974 by the GARGAMELLE experiment, a bubble chamber at CERN, which recorded neutrino-induced events explainable only by the existence of a massive neutral vector boson. These events provided the first experimental evidence for the Z^0 boson.

In 1983, another historic breakthrough was achieved with the experimental detection of the W^{+-} and Z^0 bosons by the UA1 [Arnison, 1983, 103-116], [Arnison, 1983, 398-410] and UA2 [Banner, 1983,476-485]; [Bagnaia, 1983, 130-140] experiments, both installed at the Sp$\bar{p}$S collider at CERN.

Fig. 10-2 shows an artistic rendering of the UA1 experiment. Examining the Central Detector (CD) of UA1 highlights the significant role of Charpak's invention in this discovery, which ultimately led to the awarding of the 1984 Nobel Prize in Physics to C. Rubbia and S. Van der Meer.

CD served as the experiment's tracker, offering 3D tracking capabilities (6 meters long, 2.2 meters in diameter). Its cylindrical geometry anticipated the design of future time projection chambers. CD is surrounded by the electromagnetic calorimeters (b), followed by the hadronic calorimeter (d). Particles that penetrate the iron absorbers (h), located after the hadronic calorimeter, can only be neutrinos or muons. The outermost detector is the muon tracker (e).

Neutrinos interact only via the weak force, allowing them to traverse the entire detector without producing a direct signal. Their presence is inferred solely from a transverse momentum imbalance in the event. The internal structure of the CD's wire planes is illustrated in Fig. 10-3 [Revol, 2023, 1-33].

A photograph of two horizontal layers of cathode and anode wires is shown in Fig. 10-4, with left-right electronic readouts illustrated in Fig. 10-5. These planes define drift regions situated within a horizontal magnetic field, oriented orthogonally to the beam pipes. This setup marked a "phase transition" from bubble chambers to electronic detectors, enabling the triggering of 17,000 field wires and 6,125 sense wires.

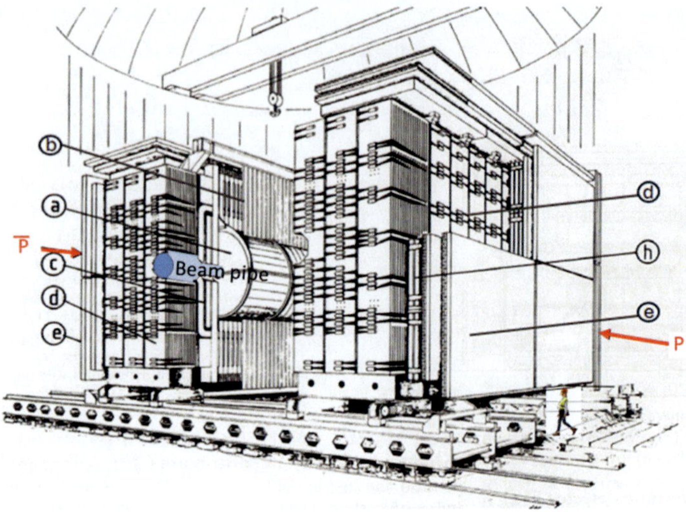

Figure 10-2. Artistic view of the UA1 experiment. Legend: (a) tracking central detector, (b) electromagnetic calorimeter, (c) magnetic field coil, (d) hadronic calorimeter, (e) drift chambers for □ detection, (h) Iron (Fe) absorber. The human silhouette at the bottom right provides a sense of the experiment's scale. Adapted from [Albajar, 1989,14-61]).

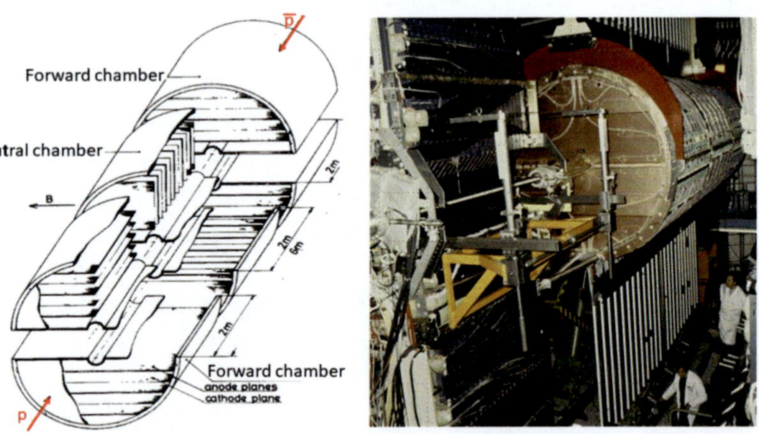

*Figure 10-3***.** *Schematic drawing and photograph of the Central Detector. Adapted from [Revol, 202, 1–33].*

Figure 10-4. Anode and cathode wire planes (CERN image).

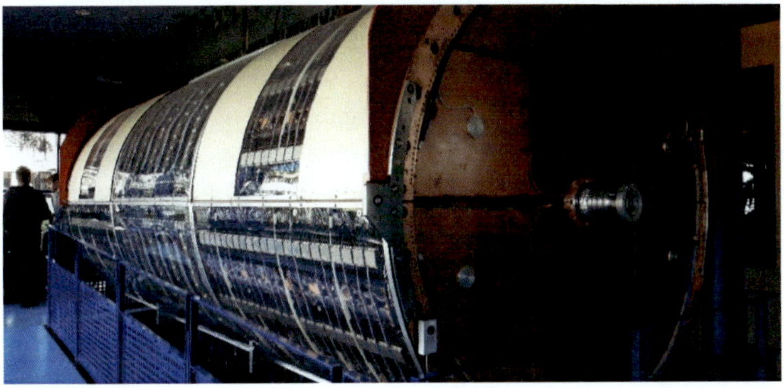

Figure. 10-5. Front-end electronics for the left-right readout of the anode wires (CERN image).

This detector configuration provided tracking information, momentum measurements, and determination of the particle's charge sign. Charpak's innovations, MWPC and the drift chamber, were fully implemented in CD. The tracking capabilities of this system were crucial in allowing physicists to interpret the complex, high-multiplicity events produced in p̄p collisions, selectively identifying electron or muon tracks as signatures of W and Z^0 decays.

The electronics also played a pivotal role in the success of the detector; 78 iterations of the wire amplifier circuit were developed "before C. Rubbia felt satisfied" (Claude Engster, UA1 electronics engineer, personal communication to the author, 2023).

As an engaging data analysis exercise, Figs. 10-6 and 10-7 show the event displays for the following two decays: a W^- decaying into e^- and ν_e, and a Z^0 decaying into μ^+ and μ^-.

In Fig. 10-6, the decay $W^- \rightarrow e^- + \nu_e$ is shown. The light blue electron track, moving downward from the interaction point, lacks a corresponding oppositely directed charged track to balance transverse momentum; therefore, an undetected neutral particle must be present. This missing particle is the electron neutrino.

Since proton–antiproton collisions have zero transverse momentum before the interaction, all final-state products must also conserve transverse momentum $p_t = 0$; however, this is not observed here. The light blue arrow opposite the electron track visually represents the missing transverse momentum, corresponding to the neutrino. By calculating the invariant mass (a conserved relativistic quantity) of the electron and the neutrino, the mass of the W boson can be determined. The decay happens almost instantaneously; thus, the electron track originates directly from the $\bar{p}$–p collision point.

The remaining orange tracks represent spectator hadrons, identified by the hadronic calorimeter surrounding the electromagnetic calorimeter and CD. This type of analysis would not have been possible without the precise tracking capability of the CD. For the first time, the experiment employed the MEGATEK online event display system, which used clusters of simple computers to emulate the powerful IBM systems (168/E, later 3081/E). This was a powerful tool for real-time monitoring of data acquisition quality.

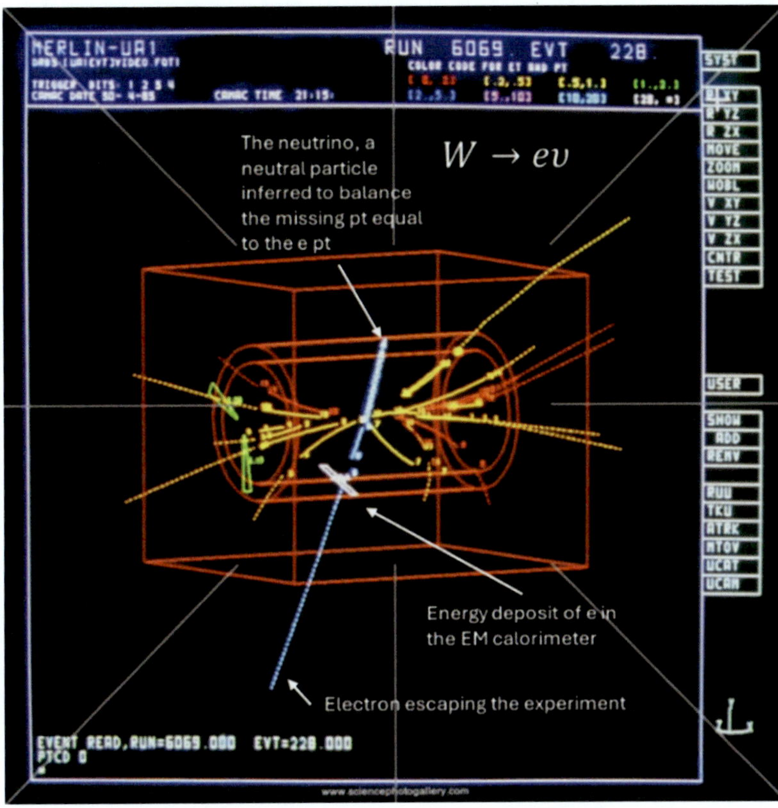

Figure 10-6. An electronic image of the W particle decay in UA1 detector at CERN. Image credit: CERN.

In the event display of Fig. 10-7, the decay $Z^0 \rightarrow \mu^+\mu^-$ is depicted. The energy deposition in the electromagnetic calorimeter, the absence of hadronic interactions in the hadron calorimeter, and the detection of signals in the muon chambers, the outermost tracking system of the experiment, together provide a clear signature identifying the muons. Calculating the invariant mass of the $\mu^+\mu^-$ pair yielded a precise measurement of the Z^0 boson mass.

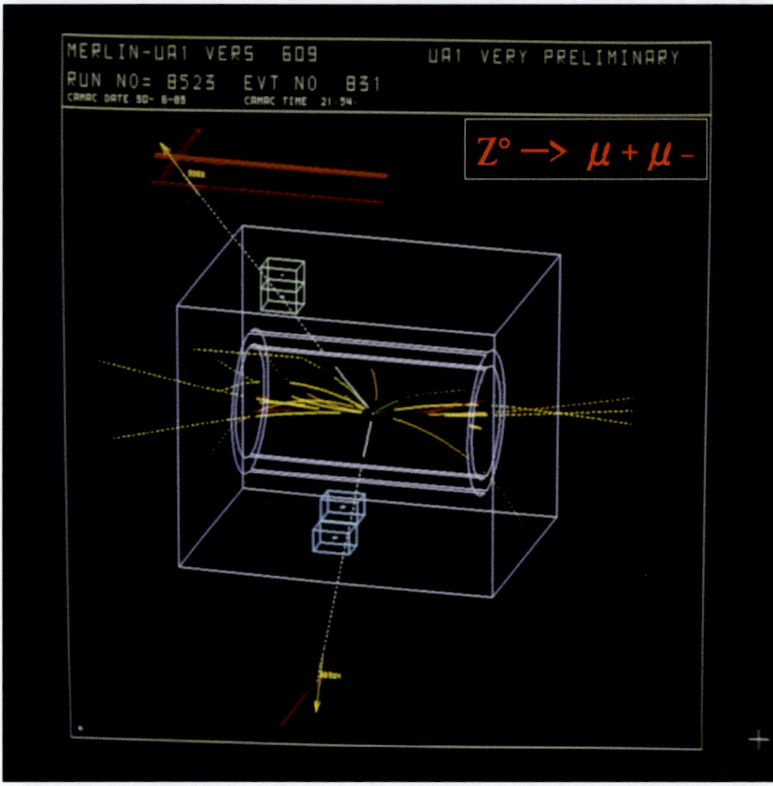

Figure 10-7. An electronic image of the Z^0 particle decay in UA1 experiment at CERN. Image credit: CERN.

10.3. For the Curious: What Antimatter Reveals in Primary Cosmic Rays

Detection of antimatter in cosmic rays, synthesized in the earliest moments of the Hot Big Bang, would represent a breakthrough of fundamental importance. Its investigation seeks to explore the observed imbalance between matter and antimatter in the present Universe, and potentially to identify sources of antimatter, if they exist.

In subnuclear interactions, scientists observe the symmetric creation of matter and antimatter. It is therefore reasonable to infer that, during a Hot Big Bang, matter and antimatter were initially generated in equal amounts.

However, the observed baryon-to-photon ratio in the Universe, approximately 10^{-9}, suggests that the original symmetry between matter and antimatter was broken by the action of three mechanisms: baryon number non-conservation, CP violation, and a departure from thermal equilibrium (e.g., due to inflation during the expansion of the Universe) [Weinberg, 1979, 850-853], [Guth, 1981, 347-356].

In everyday life, we observe ourselves and everything around us composed solely of matter. Therefore, the detection of primordial antihelium nuclei, formed from primordial antiprotons and antineutrons before annihilation, in cosmic ray experiments could provide crucial insights into the mystery of antimatter's disappearance.

Conversely, string theory does not predict observable primordial antimatter, as it allows for a Universe that may not have originated from a Hot Big Bang; instead, it would contain only primordial matter. In string theory, fundamental physical events are described within a mathematical framework of 11 dimensions, four of which correspond to the space-time of general relativity. Strings are the elementary entities from which particles emerge. In this framework, the mathematical singularity of general relativity at the Big Bang, associated with infinite energy density and equal creation of matter and antimatter, would not exist.

As discussed in earlier sections, numerous balloon-borne experiments launched since the 1980s have been designed as antimatter detectors to explore these questions. Their primary goal was to measure the abundances of antiprotons and positrons in cosmic rays.

Despite the low-precision measurements, the observed production of antiprotons and positrons appeared to exceed the levels expected from simple secondary interactions of cosmic rays with the interstellar medium (ISM) [Gaisser,1992, 174-183].

The measurement of the antiproton-to-proton ratio is shown in Fig. 10-8 [Boezio, 2005, 255-270], yet the reason for this excess remained unclear. Theorists proposed production mechanisms involving dark matter annihilation or black hole interactions as possible explanations.

Typically, balloon flights lasted about 20 hours, a duration insufficient to collect enough data on rarer events such as anti-deuterons, anti-tritium, and anti-helium, if they existed at all.

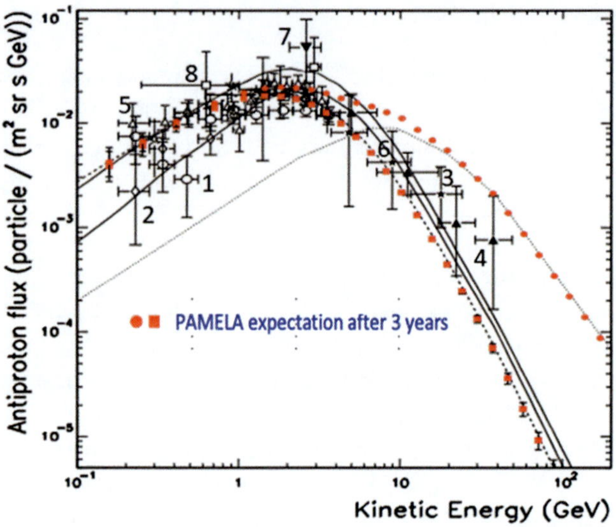

Figure 10-8. Recent experimental antiproton $\overline{\mathrm{p}}$ spectra: (1) BESS00, (2) BESS99, (3) AMS, (4) CAPRICE98, (5) BESS95+97, (6) MASS91, (7) CAPRICE94, and (8) IMAX92. Also shown are theoretical calculations for pure secondary $\overline{\mathrm{p}}$ production (solid and dashed lines) and for pure primary $\overline{\mathrm{p}}$ production (dotted line). The expected PAMELA performance is indicated for the case of a purely secondary component (solid boxes) and for a scenario including an additional primary component (dots). Only statistical errors are included in the expected PAMELA data. Adapted from [Boezio, 2005, 255–270].

To improve the sensitivity of these measurements, scientists proposed specialized satellite-based experiments capable of collecting data over several years. The Alpha Magnetic Spectrometer (AMS-02), installed on the International Space Station, stands as a prominent example. Its spectrometer measures the charge, velocity, and energy of cosmic ray particles while searching for primordial antimatter in primary cosmic rays. AMS also contributes to identifying the origins and propagation mechanisms of antimatter in the interstellar medium.

So far, AMS has detected no light antinuclei events, pushing the upper limit of their possible abundance relative to matter nuclei to about 10^{-8}.

An example of a balloon-borne experiment is TS93. Below is a more detailed description of its superconducting spectrometer. Results from the prototyping phase at the CERN PS (proton synchrotron) test beam for its TRD are presented, along with insights into the physics data analysis.

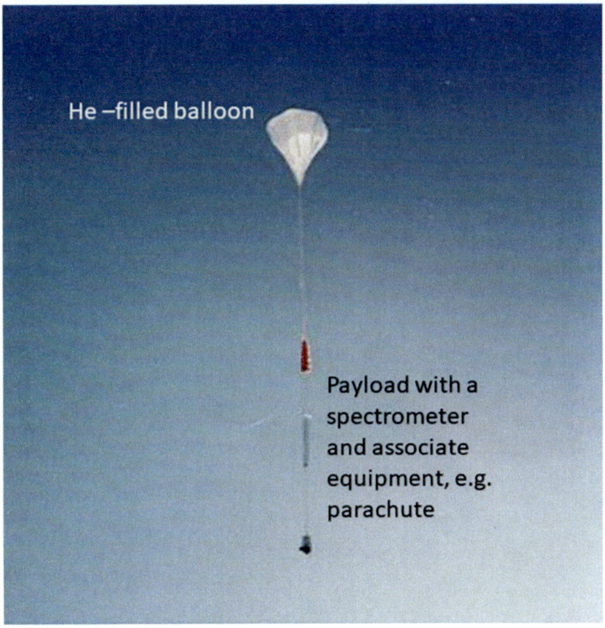

Figure 10-9. Photograph of TS93 payload made during its take-off. Adapted from http://theor.jinr.ru/~vnaumov/Eng/UHECR/Lecture01.pdf.

In Fig. 10-9, the take-off of the spectrometer is shown. Fig. 10-10 illustrates all the detectors integrated into the payload that were part of this NASA mission. The tracking system relies on MWPCs and drift chambers, placed within a magnetic field to measure particle curvature for momentum determination.

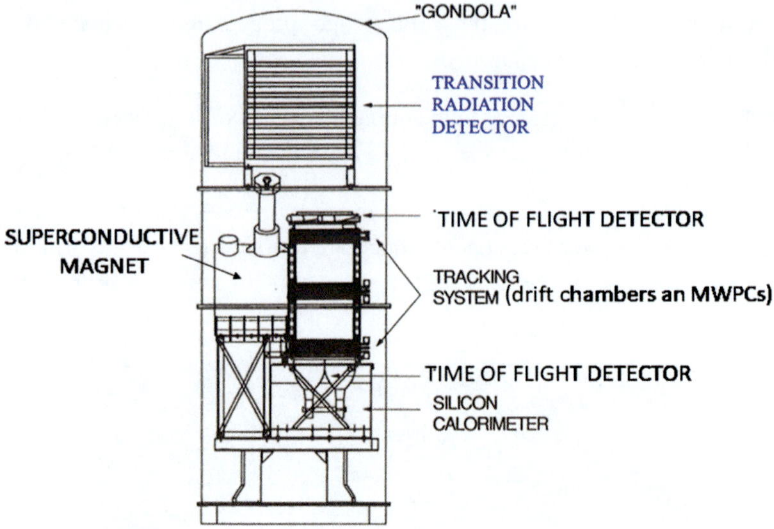

Fig. 10-10. The complete detector housed in the container (gondola). Adapted from [Barbarito, 1995, 588-600].

The spectrometer includes a 10-layer TRD (Fig. 10-11), which uses MWPCs and carbon fiber radiators, designed to distinguish between e^+/p and e^-/antiproton in the energy range of 5–60 GeV.

The TRD enables the selection of a purer positron sample from the much more abundant protons in cosmic rays. A set of plastic scintillators was installed at both the top and bottom of the spectrometer to provide the primary trigger for the experiment.

Data from the time-of-flight system were used to differentiate between particles traveling upward and those traveling downward. The pulse height data provided a measure of the particle's ionization losses and thus served as a preliminary estimate of its energy. Finally, at the base of the spectrometer, an electromagnetic imaging calorimeter was installed, composed of silicon layers (providing x-y coordinates) interleaved with tungsten layers, each one radiation length thick.

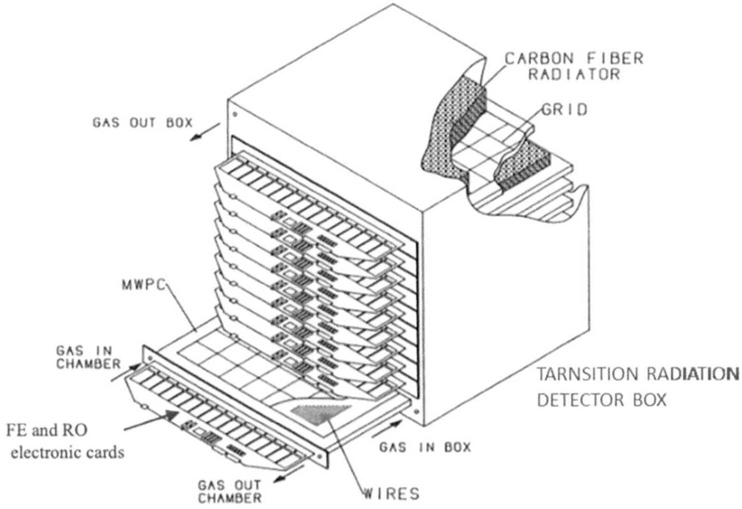

Figure 10-11. TRD artistic view with the inner MWPCs and radiators details. Adapted from [Barbarito, 1995, 588-600].

The TRD's performance was tested at the CERN PS using beams of electrons and pions with a momentum of 4 GeV/c.

In Figs. 10-12 and 10-13, the particle identification capabilities of the 10-layer TRD are shown. The pions (with $\gamma \sim 40$) did not emit transition radiation, while the electrons (with $\gamma \sim 8000$) did. This is evident in Fig. 10-12, where, on average, pions produce a smaller number of clusters (mainly due to delta rays) compared to electrons (which produce both TRD photons and delta rays). Improved separation between these two particle types can be achieved by using an additional parameter called the Likelihood Indicator (Fig. 10-13) [Barbarito, 1995, 588–600].

Applying the same identification algorithm to the flight data improved the separation of positrons from protons in the TRD, since both particles carry a positive electric charge but differ significantly in their Lorentz γ factor.

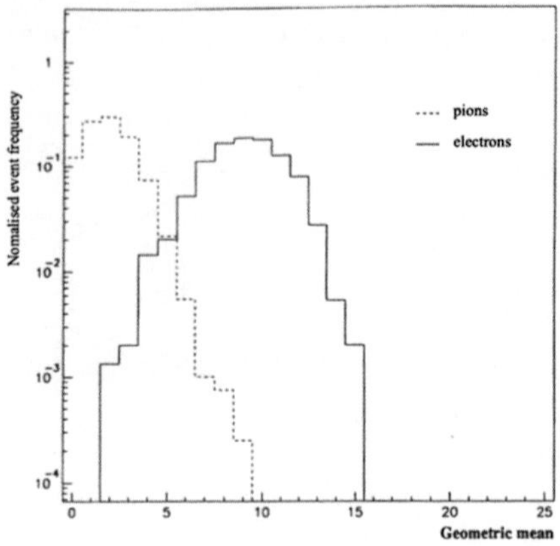

Figure 10-12. Distribution of geometric mean of the number of wire planes and the total number of clusters along the track for pions (dotted line) and electrons (solid line.) Adapted from [Barbarito, 1995, 588-600].

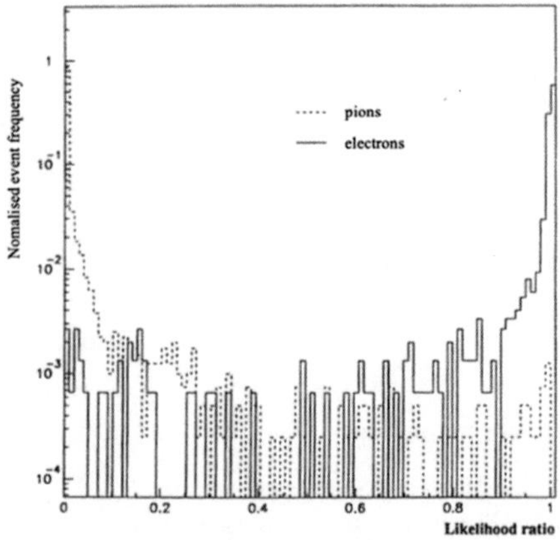

Figure 10-13. Likelihood distributions for pions (dotted line) and electrons (solid line). Adapted from [Barbarito, 1995, 588-600].

In 1996, data analysis revealed that the measured positron flux exceeded the levels expected from secondary interactions of cosmic rays with the interstellar medium. This finding suggests that a fraction of the positrons has a primary origin.

The explanation for this excess remains uncertain; however, it implies the existence of different sources of positrons in outer space. One approach to disentangling these contributions involves software simulations that model the various production mechanisms in known astrophysical sources, including possible primordial production during the Big Bang.

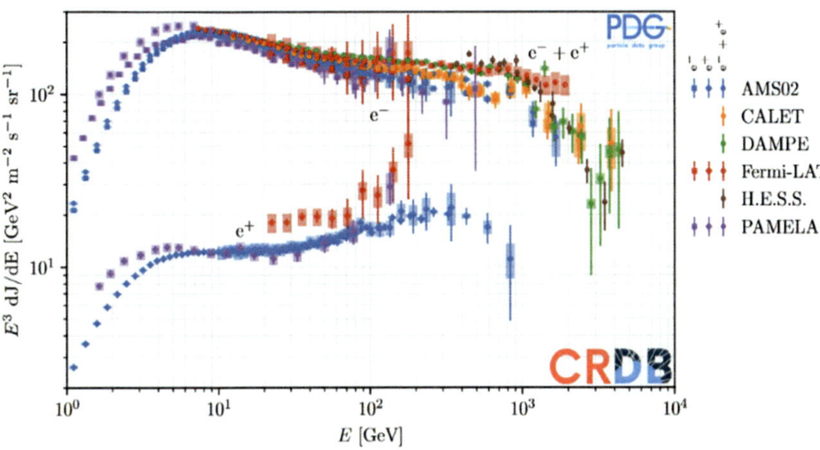

Figure 10-14. Intensities of cosmic rays electrons (squares), positrons (diamonds), and their sum (dots) as a function of the kinetic energy E. The spectra have been multiplied by E^3 to enhance the visibility of the spectral features. Data from the most recent space experiments AMS02, CALET, DAMPE, Fermi-LAT and PAMELA are included. Reproduced from [Navas, 2025, 1-2382] with permission from Particle Data Group.

The positron excess observed in TS93 was later confirmed by other experiments, (see Chapter 8). Some theoretical models attribute the excess to dark matter decay or annihilation, although direct evidence for it is still lacking; its existence is inferred only indirectly, through its gravitational effects on the rotational velocities of certain galactic arms. A summary of the current findings in this field is provided in Fig. 10-14. To date,

PAMELA and AMS have measured the rise in the positron energy spectrum near 100 GeV with the highest precision.

This line of research continues, with the hope of eventually detecting light antinuclei.

Concluding remarks

We have highlighted the substantial contribution of MWPCs and drift chambers in the identification of the charm quark, as well as the $W^{\pm}$ and Z^{0} bosons, the intermediate vector bosons of the weak force. These achievements stand as two measurement milestones within two theoretical frameworks: the quark model and the electroweak interaction. Together, they represent two fundamental pillars for the future development of the Standard Model of elementary particles and forces.

MWPCs have also played a significant role in measuring the excess of positrons and antiprotons in primary cosmic rays, helping to constrain contributions from various potential antimatter sources within our galaxy. Ultimately, the absence so far of light anti-nuclei in primary cosmic rays has yet to provide insight into the matter, antimatter imbalance in the Universe.

CHAPTER 11

MWPCs APPLICATION BEYOND HEP AND ASTROPHYSICS

Abstract

In this chapter, we briefly review the applications of MWPCs in medicine, biology, crystallography, homeland security, and plasma diagnostics. Originally developed for high-energy physics and astrophysics, MWPCs have since found diverse applications across multiple fields. Their exceptional ability to detect and track charged particles with high precision and efficiency has made them invaluable tools in a variety of scientific, medical, industrial, and research domains. Below are several notable applications of MWPCs.

Introduction

Until the 1990s, MWPCs were the only large-area electronic detectors capable of detecting single X-ray photons and charged particles with both high sensitivity and spatial resolution.

At that time, solid-state detector technology was still in its early stages of development and could not yet compete with the performance of MWPCs. As a result, MWPCs became highly attractive for many applications where single-photon counting or single charged-particle detection was essential.

A comprehensive review of MWPCs and their applications can be found in [Nappi, 2013, 1–323].

Below, a few relevant examples will be presented.

11.1. Early applications in medicine, biology, and crystallography

Historically, photographic film was the primary medium for imaging in fields such as medicine, biology, and crystallography. One of the earliest medical imaging applications of MWPCs was likely developed by a group in Novosibirsk [Baru, 1985, 165-169], who designed an X-ray scanner operating in single-photon counting mode for low-dose diagnostic examinations.

In this system, the X-ray source and the MWPC moved together vertically, detecting photons that had passed through the patient's body. The device was commercially produced and installed in more than 40 clinics across Russia.

Subsequently, a more advanced full-body scanner was developed by a team led by G. Charpak at the French company Biospace. These devices were installed in several French hospitals, where they operated successfully.

There were also efforts to apply MWPCs to medical imaging using a different approach. Instead of relying on scanning techniques, some researchers employed a two-dimensional readout method based on delay lines. In this method, the spatial coordinates of the charge avalanches were determined by measuring the time delays relative to a "start" signal induced on the anode wires.

One example of this approach was an MWPC system developed for nuclear cardiology applications [Lacy, 1988, 369-376]. In this case, a short-lived radionuclide was injected into the patient's bloodstream, and a sequence of images was captured during its transit through the body

There have also been numerous attempts to use MWPCs for the detection of beta radiation (see, for example, [Anisimov, 1980, 67-69] or [Angelini, 1988, 430-435].

Biological substances are primarily composed of carbon, oxygen, and hydrogen, with smaller amounts of nitrogen and other elements. For biological studies, radioactively labelled elements such as tritium (^3H) or carbon-14 (^{14}C) are often used, and imaging their beta emissions provides valuable information for researchers.

A notable advancement in this field was achieved by G. Charpak and his collaborators at the Biospace company, who developed and commercialized two-dimensional beta imaging detectors specifically for biological research. These detectors were based on optical chambers (see [https://biospacelab.com/]) and were installed at several universities and research institutions.

MWPCs also successfully replaced photographic film in several crystallographic studies (https://scholar.google.com/scholar?hl=fr&as_sdt=0%2C5&q=MWPC+in+crystallography&btnG=). One particularly noteworthy development was a specially designed spherical MWPC, created to minimize the parallax effect commonly encountered with conventional gaseous detectors in crystallographic measurements [Kahn, 1980,337-344].

11.2. Applications in Security devices and in muon muonography and tomography

Besides medical applications, there have been built MWPCs for security imaging: cargo, passengers in airports, etc. Below just one example will be given.

In work [Dorion, 1987, 442-448] a MWPC-based scanner is described. It obtained the images of gamma photons (100-400 keV) passed through the object to be examined. The scanner consists of the gamma source and MWPC with a tantalum converter inclined at 1 degree with respect to incoming photons. During the data taken, they moved synchronically as in the case of medical scanners. It was capable of capturing images of large objects, such as cars. The position resolution achieved was about 5 mm and the detection efficiency was 30%. This detector operated for some time in one of Paris's airports.

Conventional, stationary MWPCs are attractive for another security device: based on muon tomography. Highly penetrating muons on their pass may experience multiple scatterings due to the Coulomb interaction with nuclei of atoms. The amount of the scattering depends on the charge Z of the corresponding nucleus. If the incoming and outgoing tracks for each muon are detected, some conclusions about the inspecting object can be drawn. One of the advantages of this method is that it allows security checks to be performed without exposing operators or examined objects to dangerous radiation. By analysing the changes in energy and trajectory, a computer algorithm produces a three-dimensional mathematical map of dense items contained in the cargo or other objects.

Several companies, such as Muon Systems (Spain), Lingacom (Israel), and SAES Getters (Italy), are producing commercial devices for muon tomography, reflecting the growing demand in this field.

The most appropriate detectors for the muon trackers are large-area gaseous imaging devices and MWPCs are among them. For the inspection of objects, e.g. trucks, muon trackers are placed above and below them and detect muon track deviations. Computer-based analysis of these data allows obtaining an image of dense objects inside the cargo.

MWPCs were used also in other muon detection-bases applications. One of them is the miography of the Sakurajima volcano which is one of the world's most active volcanoes with over 3,000 explosive eruptions during the last five years (see, for example, https://indico.cern.ch/event/1033631/time table/?view=standard and references therein). The principle of volcano muography is illustrated in Fig. 11-1.

The Sakurajima Muography Observatory operates 11 MWPC-based systems with a sensitive surface area of approximately 8 square meters. Lead shielding plates, 2 cm thick and coated with 4 mm of stainless steel, are placed between the MWPCs to absorb or deflect low-energy background particles. These charged, soft particles, primarily hadrons and low-energy muons, approach from the direction of interest as a result of interactions without having penetrated the structure.

Monitoring specific subsurface magma movements provides valuable insights for volcanological research and hazard assessment. Muotomography enables the visualization of subsurface density structures within craters, helping to reveal the causes behind their activation and deactivation.

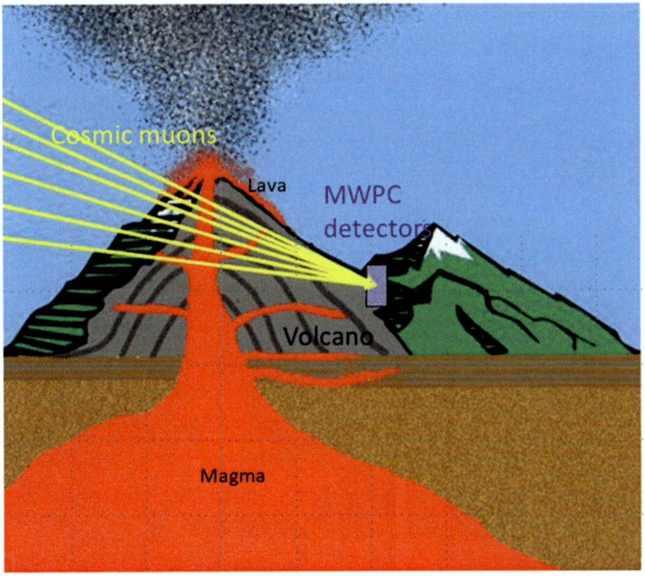

Figure 11-1. Schematic illustration of the volcano muography principle. The detector measures the attenuation of muon flux after it passes through the volcanic structure, enabling estimation of the internal structure by analysing the resulting density distribution in the muogram. Adapted from https://slideplayer.com/slide/5082833/.

11.3. Plasma diagnostics

As described in Chapter 5, the first MWPCs filled with photosensitive gases were developed simultaneously by two independent groups, one at CERN [Seguinot, 1977, 377-391] and the other at the P.L. Kapitza Laboratory in Moscow [Bogomolov 1978, 778-783]. The work of Seguinot et al. [Seguinot, 1977, pp. 377–391] is well known and has made significant contributions to the development of various RICH detectors. In contrast, the application of photosensitive MWPCs in plasma diagnostics is much less familiar, primarily because most of the related publications were originally published in Russian. The Russian group investigated stationary plasmas induced by ultrahigh-frequency microwaves (UHF, λ = 25 cm or 50 cm) in cylindrical resonators filled with H_2, D_2, or noble gases at pressures ranging from 1 to 30 atmospheres [Peskov, 1980, 1-4 10] (Figs. 11-2 and 11-3). The power dissipated in the plasma column was approximately 100 kW.

An interesting feature of this plasma was the sporadic appearance of ionization instability leading to the creation of dense short-lived (a few μs) filaments inside the plasma volume. The stationary plasma column emitted very strong molecular spectra (H_2 /D_2) and a so-called H^- continuum [Diatroptov, 1971,1038-1043]. On this huge background, it was extremely difficult to observe these short-living filaments either in the visible (380-700 nm) or UV (≈180-380 nm) regions of spectra. Thus, the most appropriate technique to visualize this instability and to detect the spectrum emitted from the filaments was the use of a photosensitive MWPC, which practically is not sensitive to the UV and visible part of the bright H^- continuum. This allowed us to perform all necessary measurements of filament parameters without interference from the background [Peskov, 1980, 1-410]. In particular, photosensitive MWPC allowed us to visualize in the vacuum ultraviolet region of spectra (104-200 nm) of filaments formation, their development, and decay. With the help of windowless MWPCs it was discovered that filaments emit soft X-ray photons [Bogomolov, 1987, 519-528].

A new method was developed to measure their spectrum. For this, the MWPC was positioned in such a way that the x-ray emission entered its volume perpendicular to the anode wires. The number of registered photons as a function of the penetration depth was measured. The obtained results were treated using the maximum-likelihood method allowing reconstruction of the X-ray spectrum from the attenuation measurements [Kosarev, 1983, 637-645], [Karabadyak, 1983, 56-64].

Similar instabilities were later observed by several other groups in various UHF discharges (see, for example, [Skorovoda, 2001,78-85]), indicating that this is a common phenomenon in plasmas generated by microwave electromagnetic fields.

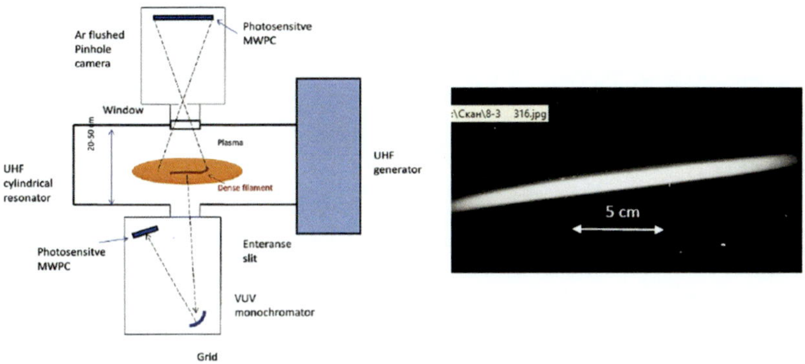

Figure 11-12. Left: *Schematic of the experimental setup for plasma studies using Multiwire Proportional Chambers (MWPCs).* Right: *Photograph of a stationary electrodeless UHF discharge in deuterium (D_2) at a pressure of 1.5 atm.*

Figure 11-3. Assembly of a high-power UHF plasma generator at the P.L. Kapitza Laboratory, Moscow.

11.4. Some ongoing R&D oriented on new potential applications

Numerous R&D efforts, both past and ongoing, have explored various potential applications of MWPCs. Below, we highlight just two examples.

11.4.1. MWPCs in PET

One of the earlier efforts in this direction was carried out at the University of California, Berkeley. It was inspired by CERN physicists Jeavons, Charpak, Townsend, and others, who developed so-called high-density avalanche chambers, devices that combined multi-wire proportional chambers (MWPCs) with a high-efficiency photon-conversion layer [Jeavons, 1975,491-503], [Townsend, 1987,1554-1562].

This technology played a pioneering role in the development of some of the earliest PET (Positron Emission Tomography) images, most notably of a mouse. High-density avalanche chambers later made significant contributions to 3D PET image reconstruction. A prototype partial-ring tomograph developed at CERN served as a forerunner of integrated PET/CT (computer tomography) systems.

Townsend later worked at the Cantonal Hospital in Geneva and subsequently in the United States, where his group was instrumental in developing the first commercial type of PET/CT scanner, a system combining functional (PET) and anatomical CT) imaging. Research and development in this area continues to this day (see, for example: Google Scholar – MWPC for PET: [https://scholar.google.com/scholar?hl=fr&as_sdt=0%2C5&q=+multiwire+chamber+for+pet&btnG=]

11.4.2. Neutron detectors

Interest in neutron detectors has increased significantly over the past decade, particularly in two key areas: instrumentation for neutron scattering science and the prevention of nuclear terrorism. Traditionally, neutron detection has relied on ^{3}He-filled MWPCs due to their low gamma sensitivity, fast response time, and relatively high spatial resolution. These advantages have contributed to a renewed and growing interest in MWPCs for neutron detection (see, for example: [https://scholar.google.com/scholar?hl=de&as_sdt=0%2C5&q=the+use+of+MWPC+neutron+detection&oq=]).

However, the so-called "^{3}He shortage crisis" in recent years has driven the development of alternative technologies. One such approach involves MWPCs coupled with solid neutron converters, such as lithium (Li) and boron carbide ($^{10}B_4C$),

See example: [Montag, 2019, 158-163] or $^{10}B_4C$ [G. Nowak et al., EPJ Web of Conference, 286, 286, 03007, 2023]. [Nowak, 2023, 1-12].

These innovations reflect an ongoing evolution in neutron detection technologies, adapting to new challenges and expanding application domains.

Concluding remarks

The versatility and precision of MWPCs make them indispensable tools across a wide range of scientific, medical, industrial, and security applications, far beyond their original role in high-energy physics and astrophysics. A remarkable example of technological spin-off!

Ongoing research and development in this field will likely reveal even more applications, continuing to drive innovation across multiple disciplines.

CHAPTER 12

CHARPAK'S FINAL CONTRIBUTIONS TO DETECTOR DEVELOPMENT

Abstract

An innovative detector, originally invented by G. Charpak and further refined to an advanced level by the Saclay group, is described. Essentially, it is a mesh parallel-plate avalanche chamber (PPAC) with a very narrow avalanche gap, less than 100 μm. This small gap introduced significant advantages: excellent position resolution, low operating voltage (approximately 500–700 V), and outstanding counting rate capabilities. The detector has since been successfully applied in several experiments.

12.1. Invention of MICROMEGAS

12.1.1 Introduction: The history of development

After his retirement, Charpak unfortunately lost access to his laboratory and technical support at CERN, which made it challenging to pursue his new ideas.

Nevertheless, Charpak, an optimistic and highly creative individual, was always able to find solutions, even in difficult circumstances.

In this case, he continued his work by pursuing three parallel initiatives:

1. He applied for an EU grant to develop a digital MWPC for low-dose X-ray cardiac imaging.
2. He founded a new company in Paris, Biospace, focused on medical and biological applications.
3. He initiated a new detector development group in Saclay, France.

Remarkably, all three efforts proved to be successful.

In 1994, one of us (VP) participated in the EU project mentioned above. Although the project was primarily focused on MWPC applications, its flexible funding policy allowed a portion of the resources to be allocated to new developments.

It was an exciting period, marked by the emergence of the first micropattern gaseous detectors, such as the microdot and microstrip. In these devices, both the anode and cathode structures were fabricated on dielectric substrates using microelectronic technology, which made it possible to reduce the gap between electrodes to as little as 100 μm, or even less [Francke, 2014,1-303].

However, these advancements came with new challenges. In such detectors, avalanches tended to develop along the dielectric surfaces, which often led to issues such as charge build-up, sparking, and limitations on counting rate. Despite these drawbacks, this was a groundbreaking new direction in gaseous detector technology, one that challenged the dominance of MWPCs. Naturally, this sparked Charpak's curiosity and determination to create something even better.

His general idea was to overcome the issues by ensuring that avalanches in microgap detectors occurred exclusively in the gas volume, well away from any dielectric surfaces. Drawing on his extensive experience with mesh-based PPACs, previously used as preamplification stages for MWPCs (see Chapter 4), Charpak decided to revisit this concept, this time with a much narrower gap, comparable to that of micropattern detectors.

At first, this seemed technically unfeasible. But guided by his vision and intuition, Georges pressed forward.

A former technician who had worked with Charpak, Bouclier, built the first prototype for us in his spare time, a small PPAC with a 0.5 mm gap (see Fig. 12-1). It performed well. However, our ultimate goal was to reduce the gap to around 100 μm, which proved to be a challenging task. The main difficulties stemmed from the quality of the mesh and the spacers.

After several attempts, Bouclier came up with an excellent solution: a specially designed flat mesh combined with precision glass spacers. This innovation allowed us to reduce the gap to 0.33 mm.

What came as a real surprise was the exceptionally low rate and small amplitude of spurious pulses. Based on prior experience with dielectric interfaces in MWPCs, we had anticipated the opposite, significant charge leakage along the spacer surfaces.

After some reflection, Georges proposed a possible explanation: as the spacer surfaces became charged, the local electric field around them diminished, thereby reducing the likelihood of charge leakage. This hypothesis was later confirmed through a 2D scan of MICROMEGAS efficiency.

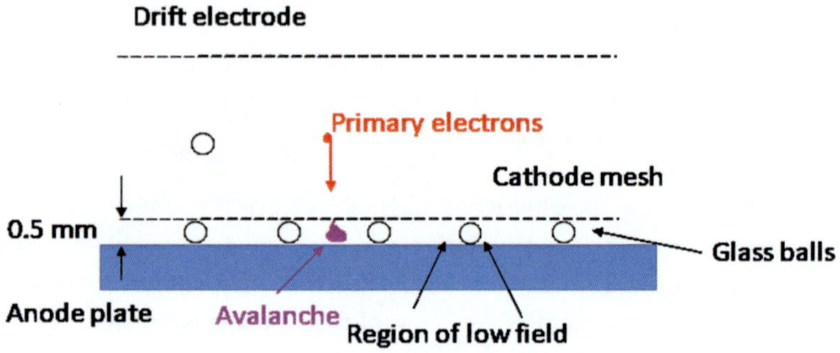

Figure 12-1. Schematic diagram of the narrow-gap PPAC prototype.

In any case, the detector worked, and Charpak was triumphant. He declared, "I'm going to call it an infinite radius wire detector." The name reflected a key design feature: a drift volume that, in principle, could be unlimited in size, funnelled toward a small region where the avalanche multiplication occurs.

Since this work did not receive adequate support at CERN, Charpak decided to move the activity to Saclay, where his collaborator I. Giomataris soon assumed leadership. Nevertheless, Charpak remained the primary driving force behind most of the developments in the new laboratory. In 1995, his group filed a patent application [Charpak, 1995, 1–30].

The team's first significant innovation was replacing the original glass ball spacers with quartz fiber spacers, which allowed the avalanche gap to be reduced to 100 μm (Fig. 12-2). Later, the quartz fibers were successfully replaced by fishing lines, an approach previously used in timing resistive plate chambers.

This design closely approached what was known as the Charpak dream. Giomataris proposed naming it a Micromesh Gaseous Structure, or MICROMEGAS. A major issue with this and earlier versions of MICROMEGAS was the bending of the mesh between spacers due to electrostatic forces, which led to a non-uniform electric field.

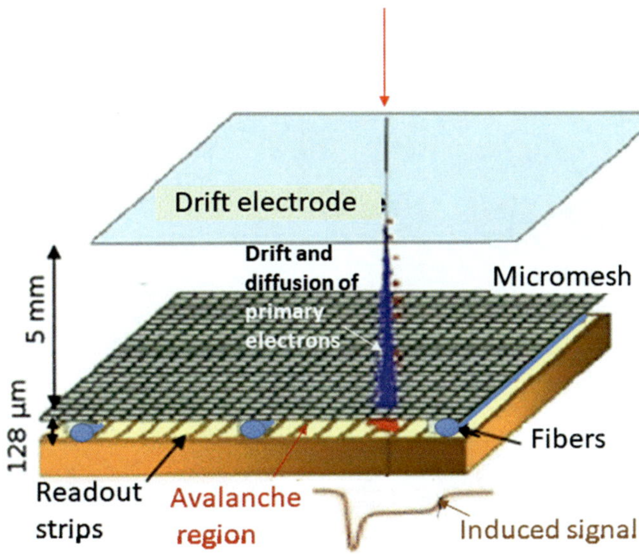

Figure 12-2. Artistic view of MICROMEGAS with spacers made of quartz fibers. The current signal from the readout strips is also shown schematically. Adapted from file:///C:/Users/Win7/Downloads/Micromegas_Detector_Using_55_Fe_X-ray_Source%20(2).pdf.

To overcome this, modern versions, often referred to as bulk MICROMEGAS, use a cathode mesh supported by a dense array of small-pitch pillars. This design greatly enhances the uniformity of the electric field within the avalanche gap (see Fig. 12-3).

Another significant advancement is that these detectors are fully fabricated using microelectronic manufacturing techniques. This represents a major technological leap, especially when compared to MWPCs, which require manual mechanical assembly by skilled technicians.

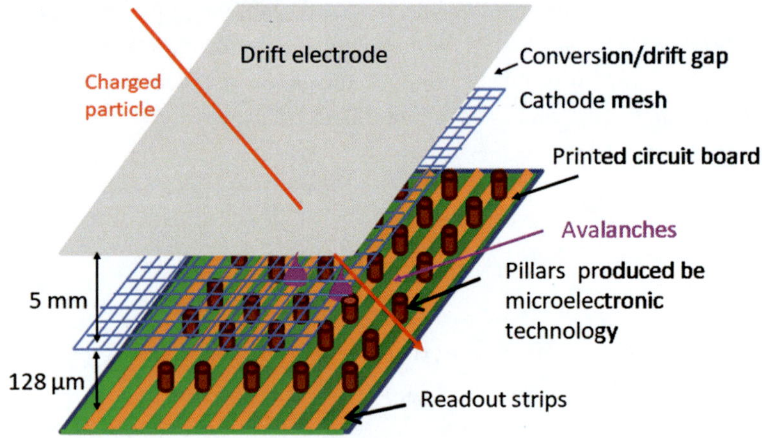

Figure 12-3. Artistic illustration of modern MICROMEGAS designs (adapted from [Kataoka, 2014, 1-9]).

In addition to improved fabrication, MICROMEGAS detectors exhibit superior performance compared to MWPCs and other micropattern detectors. Advantages include low operating voltage (see next pragraph), better time, energy, and spatial resolutions, and the ability to function at extremely high counting rates.

12.1.2. The main characteristics of MICROMEGAS

Fig. 12-4 shows the gas gain as a function of applied voltage. These measurements were performed in an Ar + 5% isobutane mixture. In other gas mixtures, MICROMEGAS operates at similar voltages, typically below

700 V. This is in stark contrast to the operating voltages of MWPCs, which generally range from 3000 to 5000 V (see Fig. 3-2).

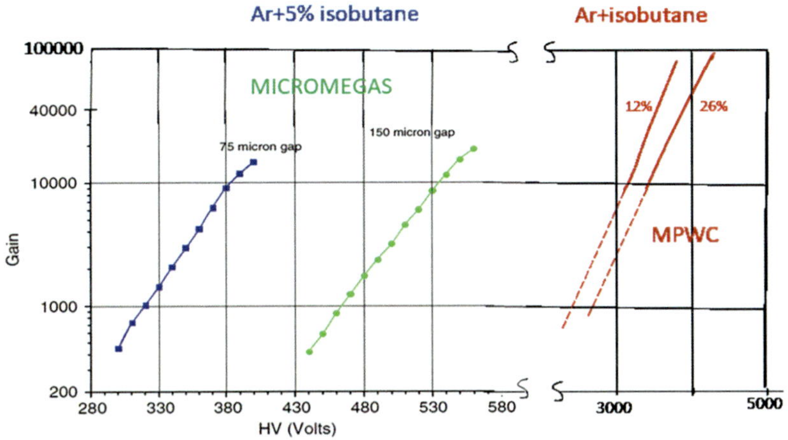

Figure 12-4. Gain curves measured from two MICROMEGAS detector prototypes (adapted from [Giomataris, 2006, 405-408] and [Bouclier, 1970, 149-161].

As shown in Fig. 12-4, gas gains up to $2 10^4$ were achievable using 6 keV photons. However, when using highly ionizing particles, such as alpha particles, the maximum attainable gain was only about 10^2. Why is there such a difference?

This is an interesting topic worth exploring.

As described in Chapter 4, the maximum achievable gain A_{max} in MWPCs is limited by feedback processes and is determined by the condition:

$$A_{max}\gamma=1$$

However, the situation is quite different in PPACs, operating in quenched gases. As shown by Raether [Raether,1964, 1-191], in PPACs with large gaps (on the order of 1 cm or more), the maximum gain is instead governed by the critical avalanche charge Q_{crit}. When this charge is exceeded, avalanches transition into streamers, often resulting in sparks.

In this case, the maximum gain is given by:

$$n_0 A_{max} = Q_{crit} \quad (12\text{-}1)$$

Here, n_0 is the number of primary electrons, and Q_{crit} depends on both the gap width and the gas used. Typically, its order of magnitude is around 10^8 electrons.

In a series of studies (see, for example, [Fonte et al., 1999, 321–325], [Ivaniouchenkov, 1999, 258–262]), it was experimentally demonstrated that the concept of the Raether limit applies to nearly all micropattern detectors, including MICROMEGAS. In the latter case, the critical charge (Q_{crit}) is approximately 10^7 electrons, which imposes a fundamental limit on the maximum achievable gain.

Another phenomenon related to the Raether limit is the rate capability of MICROMEGAS, that is, the maximum counting rate at which the detector can operate reliably and without sparking.

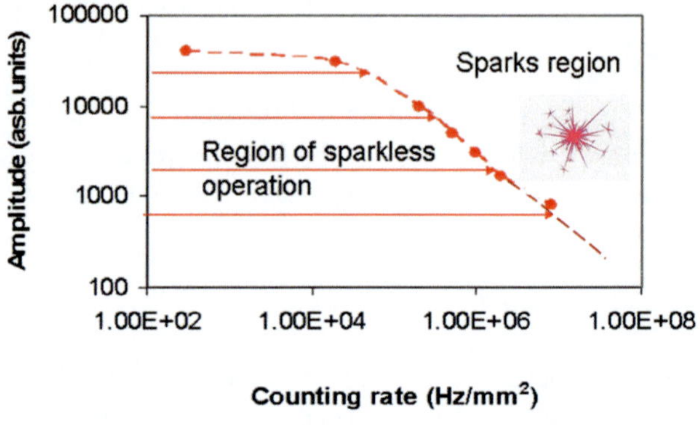

Figure 12-5. Signal amplitude (arbitrary units) versus counting rate for a MICROMEGAS detector irradiated with 6 keV photons (adapted from [Francke, 2014,1-303]).

In MICROMEGAS, as in PPACs, the signal amplitude remains constant with increasing counting rate. However, the maximum achievable gain (A_{max}) before sparking occurs depends on the signal amplitude, as illustrated in Fig. 12-5. This behaviour is notably different from that of MWPCs, where the signal amplitude begins to decrease beyond a certain counting rate due to the space charge effect (see Fig. 3-8 in Chapter 3).

As demonstrated in [Peskov,2009, 1-44], this effect is associated with the spatial and temporal overlap of avalanches. For a given gas gain $A<A_{max}$, the cumulative charge Q_{crit} can be reached due to overlapping avalanches, leading to streamer and spark formation. Calculations based on this assumption accurately reproduce the curve shown in Fig. 12-5, which delineates the boundary between the "sparkless operation" and the "spark region."

Another factor contributing to breakdowns at high counting rates may be a form of the Malter effect: jets of electrons emitted from the cathode under ion bombardment (see [Iacobaeus,2002, 1622-1628] for more details).

In MICROMEGAS detectors, due to the small amplification gap, the region where induced charges accumulate on the anode and cathode electrodes, caused by the movement of avalanche electrons and positive ions, is very narrow, typically a few hundred micrometers. The position resolution is significantly better than this and depends on the readout geometry as well as the electronics and reconstruction algorithms used. For instance, in early studies [Alexopoulos, 2009, 1-12], a spatial resolution of approximately 67 μm was achieved with a single MICROMEGAS module.

In another experiment using a MICROMEGAS telescope, a position resolution of approximately 14 μm was achieved at normal incidence of a pion beam [Derre, 2002, 523-531].

Similar to MWPCs, the time resolution of MICROMEGAS detectors is determined by the jitter in the position of primary electron clusters created by ionizing radiation within the drift volume. Consequently, it depends on various factors such as gas composition, drift velocity, and others.

Due to these limitations, the time resolution of MICROMEGAS is not significantly better than that of MWPCs. However, there is potential for improvement. For example, incorporating preamplification in the drift region, as proposed in [Peskov, 2001, 1070-1074], can enhance performance. Such a detector can be regarded as a hybrid between MICROMEGAS and PPAC, the latter being well known for its excellent time resolution of about 1 ns or even better in thin-gap designs.

To evaluate the intrinsic resolution of MICROMEGAS, one must minimize the jitter in the creation of primary electrons. As demonstrated in earlier work [Charpak, 1991, 63-68], this can be achieved using a UV-sensitive detector combined with a solid photocathode (see Fig. 12-6).

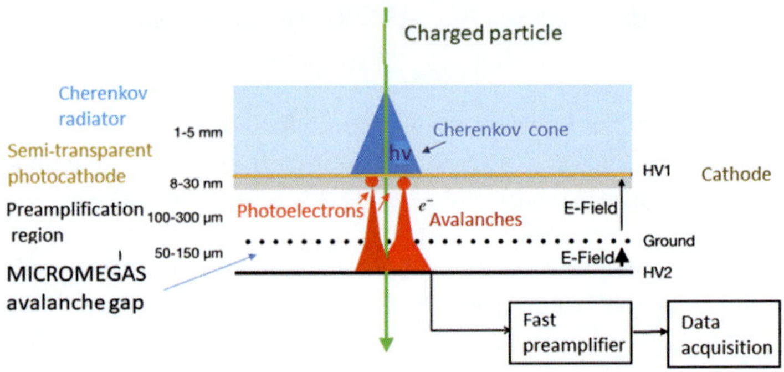

Figure 12-6. Schematic view of a MICROMEGAS detector combined with a solid photocathode. Adapted from https://www.mdpi.com/2076-3417/11/12/5362.

When the photocathode is irradiated by a very short UV pulse (e.g., Cherenkov light), photoelectrons are emitted and, under the influence of the drift field, begin to move toward the cathode mesh. Their arrival time is primarily influenced by diffusion, which helps minimize overall timing jitter.

The time resolution of such a detector naturally depends on several factors, including the gas mixture and drift gap, the electronics, gas gain, the number of primary photoelectrons emitted from the photocathode, and more. Under various experimental conditions and configurations, the measured time resolution has ranged from 20 to 50 ps, comparable to that of small-gap resistive parallel-plate chambers (see, for example, [Abbrecia, 2018, 1-379]). In the literature, one can find a range of reported values for the energy resolution of MICROMEGAS detectors, which may lead to some confusion.

Using a well-collimated X-ray beam to irradiate only a very small portion of the detector's sensitive area, specifically the region between spacers where the electric field is uniform, an energy resolution of approximately 12% FWHM for 6 keV photons has been measured.

However, in practical experiments, the area exposed to radiation is typically much larger than in these controlled laboratory setups. As a result, due to electric field nonuniformities across the larger area, the energy resolution degrades and typically reaches values around 18%, which is comparable to that of MWPCs.

As demonstrated by the data presented in this paragraph, MICROMEGAS exhibits superior rate capabilities, as well as better time and spatial resolutions compared to MWPCs. However, like any device, it also has its drawbacks. The most significant of these is its high probability of sparking in the presence of heavily ionizing particles, a fundamental issue related to the Raether limit. Such sparks can damage both the front-end electronics and the MICROMEGAS detector itself.

This problem has been addressed in a modified design: MICROMEGAS with resistive electrodes.

12.1.3. MICROMEGAS with resistive electrodes

The first prototype of a MICROMEGAS detector with resistive electrodes was described in the work by R. Oliveira et al. [Oliveira, 2010, 3744-3752]. A schematic of the design is shown in Fig. 12-7. This design is similar to that of a standard bulk MICROMEGAS; however, the cathode mesh is made of resistive material instead of the conventional metallic one. The detector operates like a typical MICROMEGAS, but in the event of sparks, which may occur at high gas gains, the resistive mesh limits the current, rendering the sparks harmless.

Another approach was proposed by T. Alexopoulos et al. [Aexopoulos, 2011, 110-118]. Their design was based on the earlier work by Fonte et al. [Fonte, 1999, 154-159], which featured a metallic cathode mesh combined with a resistive anode (Fig. 12-8). The key innovation introduced by Alexopoulos et al. was the use of separate resistive strips instead of a continuous resistive layer. This modification was designed to prevent charge from spreading across multiple readout strips and to localize the area affected by a discharge. Several variations of this concept were later developed (see, for example, [Chefdeville, 2021, 1–12].

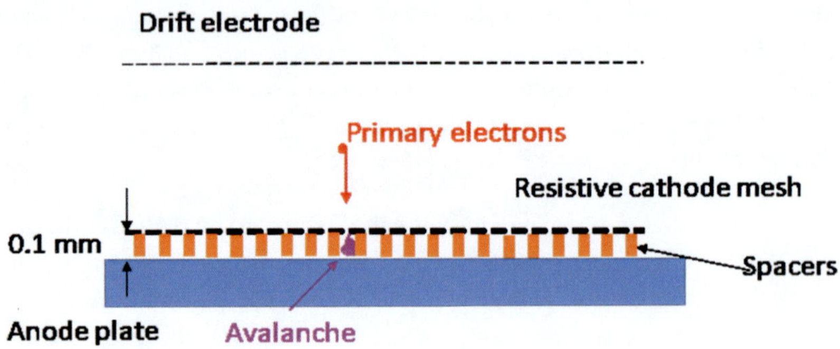

Figure 12-7. Schematic view of a bulk MICROMEGAS with a resistive cathode mesh.

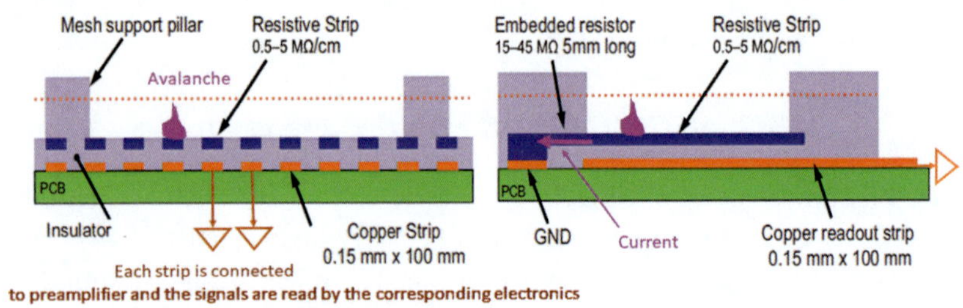

Figure 12-8. Schematic illustration of the detector principle (not to scale), showing the resistive protection scheme: (left) view along the strip direction; (right) side view, perpendicular to the strip direction (adapted from [Aexopoulos, 2011, 110-118).

Thanks to their enhanced reliability, resistive MICROMEGAS detectors open up new possibilities for applications, some of which will be discussed in the following paragraph.

12.1.4. MICROMEGAS applications

MICROMEGAS has already been used in several small- and large-scale physics experiments, such as serving as a tracker in CLAS12 (Deeply Virtual Compton Scattering experiment at Jefferson Lab) and COMPASS (Common Muon and Proton Apparatus for Structure and Spectroscopy), in TPC detectors, and more. Notable applications of resistive MICROMEGAS include their use in ATLAS (A Large Ion Collider Experiment), T2K (Tokai to Kamioka), CLAS12, and the Pyramid Muon Tomography project. A detailed review of these applications can be found in several publications (see, for example, [Attie, 2021, 1-35]).

Here, we highlight the application of resistive MICROMEGAS in the ATLAS small wheel, where two of Charpak's technologies, thin-gap MWPCs with resistive cathodes and resistive MICROMEGAS, are combined into a powerful detector system that provides both trigger and tracking information. Geometrically, the small wheel is a disc approximately 10 meters in diameter, consisting of layered sandwiches of wedge-shaped quadruple thin-gap MWPCs and MICROMEGAS detectors.

12.1.5. Concluding Remarks

Thanks to the dedicated efforts of the Saclay team, Charpak's concept of an infinite-radius gaseous detector, later named MICROMEGAS, was successfully realized and developed into a practical device. MICROMEGAS can be considered a descendant of the PPAC.

A major advancement was the introduction of the resistive electrode version of MICROMEGAS, which provided spark protection and significantly expanded its range of applications.

For more information on the further development of MICROMEGAS and other micropattern detectors, see the following references: Francke, T., & Peskov, V. (2014). Innovative Applications and Developments of Micro-Pattern Gaseous Detectors. Engineering Science Reference, 1–303.

Sauli, F. (2021). Micro-Pattern Gaseous Detectors. World Scientific, 1–349.

12.2. George's final research projects

Abstract

This text explores the innovative contributions of Georges Charpak to environmental projects. His work bridged the gap between fundamental research and societal needs, demonstrating the practical impact of scientific advancements on addressing environmental challenges.

Introduction

Between 2006 and 2010, one of Georges Charpak's final research projects focused on the development of environmental and health alarm devices using gaseous detectors operating in pulse counting mode. Charpak dedicated significant effort to this work, which encompassed the following specific areas of research and development:

- Development of low-cost gaseous detectors for large-scale online monitoring of radon (Rn) in seismically active regions.
- Creation of sensors for detecting hazardous and flammable gases.
- Design, testing, and optimization of devices for early forest fire detection.
- Manufacturing of specialized high-efficiency solar-blind photocathodes.

This work was carried out as part of a collaboration between CERN and the École des Mines in Saint-Étienne, France.

While environmental alarm, monitoring, and health safety devices operating in air ionization mode are already commercially available, such as smoke detectors, hazardous gas sensors, medical X-ray devices, and charged particle dosimeters, their sensitivity is largely determined by the performance of compact picoammeters integrated into the devices, typically operating in the picoampere (pA) range.

In this context, the collaboration between CERN and the École des Mines in Saint-Étienne, led by Georges Charpak, explored the use of gaseous detectors operating in avalanche mode. This approach enabled a high signal-to-noise ratio and reduced sensitivity to electromagnetic interference and low-frequency acoustic vibrations. In such a configuration, the gaseous detector acts as a naturally integrated preamplifier for the electronic circuitry, thereby enhancing overall sensitivity.

Additionally, the collaboration investigated the feasibility of achieving avalanche multiplication directly in air. This innovation simplified the sensor design and significantly reduced production costs.

12.2.1. Low-cost gaseous detectors for massive online monitoring of Rn in seismic active regions

Some studies have suggested the potential to correlate elevated concentrations of Rn gas in soil or groundwater, or changes in its concentration, with the early prediction of earthquakes. One of the most striking observations comes from satellite data, which showed an increase in infrared radiation several days before the magnitude 9 Tohoku Earthquake in Japan (https://arxiv.org/ftp/arxiv/papers/1105/1105.2841.pdf). This phenomenon was attributed to the release of radon caused by minor movements of the Earth's crust preceding the quake.

Researchers propose that the released radon ionizes particles in the air, leading to the condensation of water molecules from their vapour state. This condensation process releases latent heat, resulting in a localized increase in atmospheric temperature.

To verify such observations on a more robust statistical basis, it is necessary to establish a wide network of inexpensive, compact, and high-sensitivity Rn detectors, deployed at key locations where earthquakes are likely to occur (fig. 12-9). While excellent commercial detectors such as the ATMOS 12dpx or RADIM3 are available, they are too costly to be deployed in large numbers. Their high price is justified by their outstanding performance, particularly in terms of spectroscopic resolution.

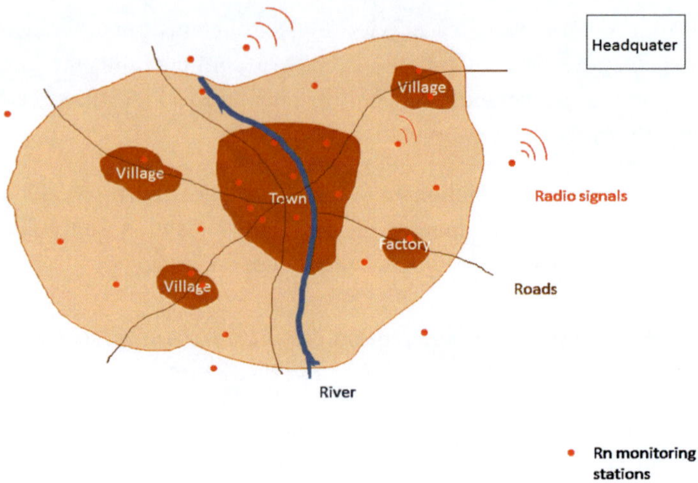

Figure 12-9. Schematic representation of potential locations for radon monitoring stations in a seismic region, equipped with radio transmitters for sending signals to a central headquarters.

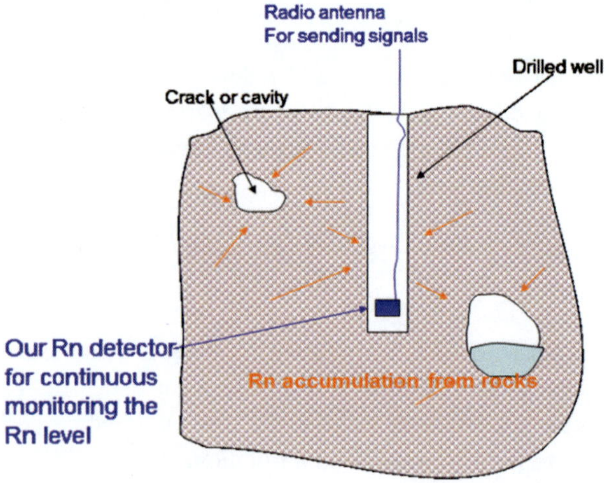

Figure 12-10. Possible arrangement of a Rn monitoring station, installed either in specially drilled wells or in the basements of residential buildings.

However, for the purpose of monitoring Rn concentrations across multiple sites and analyzing correlations between these measurements, such high energy resolution is not always essential. In most cases, it is sufficient to detect signals that exceed a predefined threshold, which can reliably indicate the presence and accumulation of radon (see Fig. 12-9).

Each monitoring station should be installed in specially designated wells (see Fig. 12-10), where radon may be released prior to an earthquake due to minor movements of the Earth's crust. A key requirement for implementing this approach is the use of low-cost, low-power-consuming sensors. The detector array must be integrated with a network of radio transmitters that send signals to a central headquarters, where the data are stored and analysed.

To extend battery life, measurements should be performed periodically, for example, for a few minutes each hour.

Two types of gaseous detectors, tailored for this specific application, were developed and preliminarily tested by the CERN–Saint-Étienne collaboration (see: https://indico.cern.ch/event/51276/contributions/2034269/attachments/967140/1373486/3_Peskov.pdf):

1. Specially designed wire-type detectors (a single-wire counter and a MWPC), capable of operating under harsh environmental conditions [Charpak, 2008, 1-18], [Charpak, 2011,187-189],
2. Micropattern detectors: MICROMEGAS and GEM, with and without resistive electrodes [Charpak, 2008, 1657-1663].

Each of these laboratory prototypes had its own advantages and disadvantages. For example, the single-wire counter featured the simplest design and the lowest cost (approximately €100, including power supply and electronics). However, its energy resolution was limited, as not all alpha particle tracks were fully confined within its relatively small volume (250 cm^3). The MWPC, on the other hand, provided significantly better energy resolution but had a more complex design, resulting in a higher cost. Micropattern detectors offer the simplest design from the standpoint of large-scale manufacturing, but they are more sensitive to environmental factors such as humidity and dust.

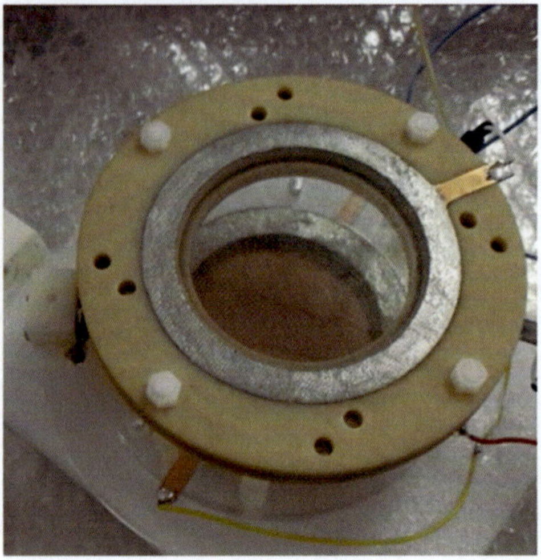

Figure 12-11. One of the design of MWPC for Rn detection, capable operating in avalanche mode in air and under harsh environmental conditions.

As a result of extensive testing, it was concluded that wire-type detectors, particularly the MWPC (see Fig. 12-11), are well-suited for large-scale Rn measurements. These detectors provide a minimum detectable activity (MDA), defined as the lowest activity level that can be reliably detected under specific measurement conditions, comparable to that of the best commercial sensors (see Table 12-1), but at nearly 100 times lower cost. Importantly, they can operate reliably in harsh conditions typical of well environments, including 100% humidity and high levels of dust.

Time of counting (min)	Single -wire proportional counters MDA	Single -wire ionization chamber MDA	MWPC MDA	**Atmos MDA** (the best on the market)
0.2	1300	6830	625	
1	420	1960	**140**	**150**
2	270	1200	75	
4	175	760	43	

Tale 12-1. Minimum Detectable Activity (MDA) for Selected Radon Detectors (source: https://indico.cern.ch/event/51276/contributions/2034269/attachments/967140/1373486/3_Peskov.pdf).

Fig. 12-12 shows the results of online Rn measurements using an MWPC as it was moved to the basement. As can be seen, even small traces of Rn were reliably detected. These features make the detector suitable for large-scale applications, such as continuous Rn monitoring for potential earthquake prediction.

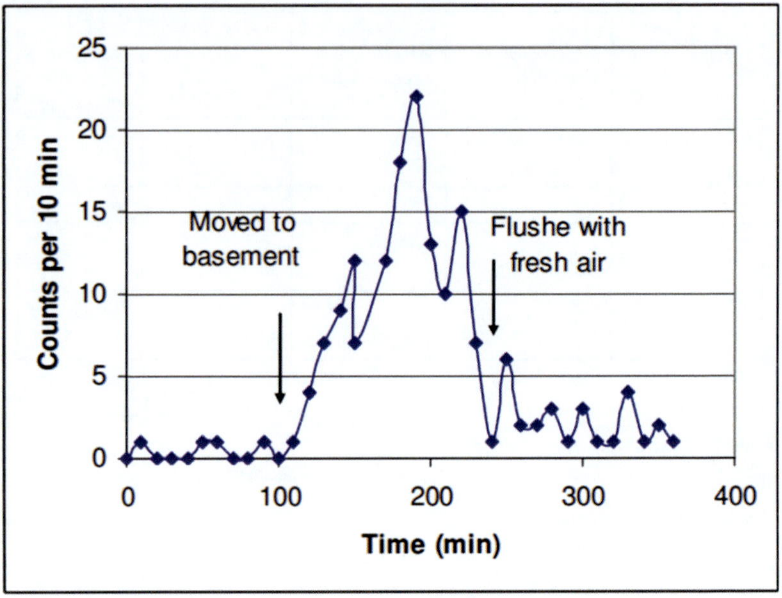

Figure 12-12. Results of the detection a small concentration of Rn in a basement with a MWPC. In the time interval of 0-100 min the chamber operated in fresh air. At t=100 min it was moved in to the basement. At t=240 min MWPC was flushed with fresh air and removed from the basement (from https://indico.cern.ch/event/51276/contributions/2034269/attachments/967140/1373486/3_Peskov.pdf).

12.2.2. Sensors for detecting hazardous and flammable gases

Among commercial sensors for detecting hazardous gases, photoionization detectors (PINDs) hold a distinctive position. These devices function as ionization chambers operating in air and are paired with ultraviolet (UV) lamps that emit photons continuously, referred to here as "continuous lamps", with photon energies ranging from 9.5 to 11.7 eV. While photons in this energy range cannot ionize air, they can ionize gases with ionization potentials (E_i) lower than the photon energy (E_v). When such gases enter the ionization chamber, they are photoionized by the UV radiation, producing electron-ion pairs. These pairs generate a measurable current when subjected to an electric field.

By employing continuous UV lamps with different photon energies, it is possible to roughly identify the types of gases present. The sensitivity of this method, given a fixed lamp intensity and detector volume, is limited by the sensitivity of the picoammeter used for current detection. At the time of the original study made by CERN-Sant- Etienne collaboration, a typical commercial PIND ionization chamber could detect gas concentrations as low as ~1 ppm for gases with low ionization potentials.

In contrast to commercial sensors, PIND detectors, developed by Charpak-led team operated in avalanche mode in air [Charpak, 2008, 1657-1663], [Charpak, 2011,187-189], where primary electrons are multiplied, resulting in significantly enhanced sensitivity.

As an example, Fig. 12-13 presents the results of measurements of various benzene concentrations in air. As can be seen, benzene at a concentration as low as 0.03 ppm was detectable in avalanche mode.

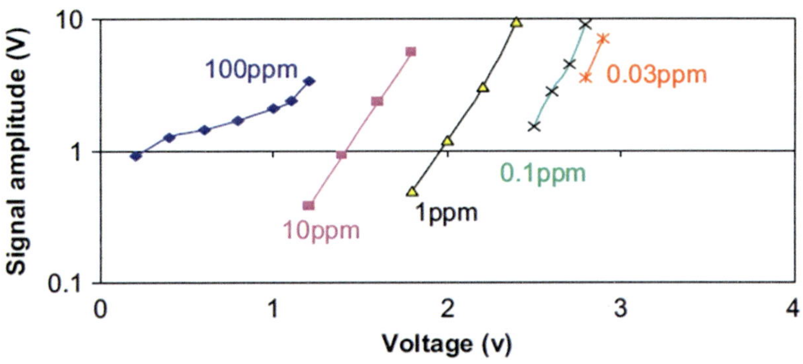

Figure 12-13. Signal versus voltage curves for a single-wire detector operating in pulse mode while detecting various concentrations of benzene. At a benzene concentration of $C_b = 100$ ppm, measurements were performed either in ionization chamber mode or at a gain of 3. At $C_b \approx 0.03$ ppm, the gas gain during measurements was increased to 4×10^3 (reproduced from [Charpak, 2011, 187–189] with permission from Elsevier).

Similar results were obtained with GEM-type detectors [Charpak, 2008, pp. 1657–1663].

12.2.3. Design, tests and optimization of devices for early forest fire detection

Each year, violent forest and bushfires in southern Europe, Australia, the western United States, and other regions around the world threaten human lives, devastate industrial and agricultural infrastructures, and inflict severe environmental damage.

Unfortunately, systematic observations indicate that the probability of forest and field fires is steadily increasing due to climate change and human activities. Considering the worldwide scale and severity of forest fire destruction, this issue must be recognized as a global disaster.

The most efficient and effective strategy for combating forest fires is early detection. Currently, several techniques are used for early fire detection, which can be broadly divided into two main categories:

1. Visual smoke and flame detection, and
2. Infrared (IR) flame detection.

Visual smoke detection is carried out from watchtowers, unmanned aerial vehicles ("drones"), and even satellites. IR detectors are typically installed on small patrol airplanes (see https://indico.cern.ch/event/50396/contributions/2023496/attachments/963749/1368395/Peskov_talk_in_Greec_final.pdf).

Other methods are also under development, including laser light scattering on smoke and local temperature monitoring with heat detectors. However, these methods are expensive and have limited capabilities. Furthermore, since most current systems operate in automatic monitoring mode, they generate a substantial number of false alarms.

To mitigate false alarms, a combination of multiple detectors is often used in industrial applications, such as fire monitoring in confined areas, to distinguish true fire events from disturbances caused by welding, lightning, etc. For instance, systems combining IR and UV detectors require simultaneous activation of both detectors before issuing an alarm.

It's worth noting that commercial UV flame detectors are available, with the most sensitive classified as EN 54-10 Class-1. However, their sensitivity is insufficient for early forest fire detection. Moreover, these detectors operate in digital mode and cannot distinguish between isolated photons or small sparks.

In contrast, the UV flame detectors developed through the CERN–St-Étienne collaboration operate on a fundamentally different principle. They use a photosensitive element based on either:

- Photoionization of vapours with low ionization potentials (such as EF or TMAE, see paragraphs 4.2 and 5.3), or
- Highly efficient solid photocathodes, e.g. CsI.

This approach offers several advantages, the most significant being exceptionally high sensitivity. Additionally, these detectors operate in proportional mode, enhancing their performance.

Fig. 12-14 illustrates the results of detecting a wooden fire (approximately $1.5 \times 1.5 \times 1.5$ m^3 in size) at various distances on a sunny day. The tests were conducted simultaneously using three detectors:

- a GEM-type detector (referred to as S-RETGEM) filled with EF vapour,
- a TMAE-filled single-wire counter, and
- a Hamamatsu R2868 Uvitron (a Class-1 detector).

In the figure, open symbols indicate the calculated counting rate versus distance, assuming an inverse-square law dependence ($1/R^2$). It is evident that both the S-RETGEM and the TMAE-filled counter effectively detect the flame at distances of up to 800 meters (the maximum distance available for these measurements). They are approximately 100 times and 1000 times more sensitive, respectively, than the Class-1 detector.

Additionally, it can be observed that the counting rate deviates from the expected 1/R² dependence, suggesting the presence of UV absorption in air.

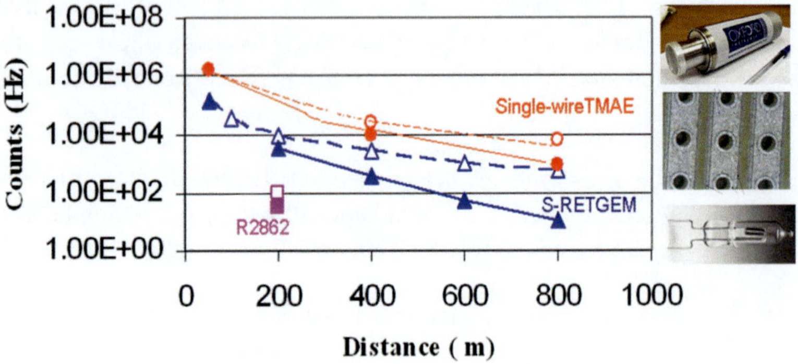

Figure 12-14. Counting rate versus distance measured with three detectors: S-RETGEM (filled triangles), single-wire counter filled with TMAE vapour (filled circles), and Hamamatsu R2868 (filled squares), in response to a 1.5 × 1.5 × 1.5 m³ flame under sunny conditions. Open symbols represent the calculated counting rates versus distance, assuming an inverse square (1/R²) dependence. The right side of the figure shows images of the detectors used, from top to bottom: single-wire counter with TMAE vapour, S-RETGEM, and Hamamatsu R2868. Adapted from [Charpak, 2009, 1–16].

UV-sensitive detectors can be installed on towers in high-risk areas, including bushes and open spaces, and are optimally used in combination with other types of flame detectors, such as CCD, IR sensor, and others. The most advantageous application is their joint operation with IR detectors on drones, as atmospheric transmission for UV radiation improves significantly with altitude. This enables these detectors to identify small fires at distances of several km. Additionally, these detectors are well-suited for a wide range of industrial applications, from factories to aircraft hangars.

12.2.4. Advanced Manufacturing of High-Efficiency Solar-Blind Photocathodes

The operation of detectors filled with photosensitive vapours becomes peculiar at temperatures below 5–10 °C, when the vapours begin to condense and form a photosensitive layer on the cathode. As a result, while the detector remains stable and sensitive to UV light, its quantum efficiency is reduced (see [Peskov, 1988, 149–160], [Charpak, 1989, 537–546], [Seguinot, 1990, 133–147]). Although maintaining the detector with photosensitive vapours at an elevated temperature is not a major technical challenge, it still introduces certain complications.

For this reason, the CERN-St-Étienne collaboration also explored the use of solid photocathodes, which exhibit stable quantum efficiency over a wide temperature range.

Among the photocathodes tested, the most promising so far has been CsI, due to its high efficiency and excellent temperature stability. This material can be exposed to air for 5–10 minutes, if necessary, which simplifies its handling. A flame detector equipped with this photocathode performs well indoors. However, for outdoor applications, it requires filters to suppress the extremely strong long-wavelength radiation from the Sun, which increases the detector's cost.

One of the approaches tested by the CERN-St-Étienne collaboration involved suppressing sensitivity at longer wavelengths by coating the CsI surface with a thin film of an appropriate material. A photoelectron escaping the CsI surface has a kinetic energy E_k, given by:

$$E_k = h\nu - \phi \tag{12-2}$$

where ϕ is the electron affinity of CsI and $h\nu$ is the photon energy. When the CsI surface is coated with a film, this layer serves as an additional barrier for the photoelectrons before they can enter the gas within the detector (see Fig. 12-15).

For a given film thickness, the larger the kinetic energy E_k, the higher the probability that the photoelectron will pass through the film. Therefore, by properly optimizing the film material and thickness, it is possible to reduce the electron yield at long wavelengths while preserving its value at short wavelengths.

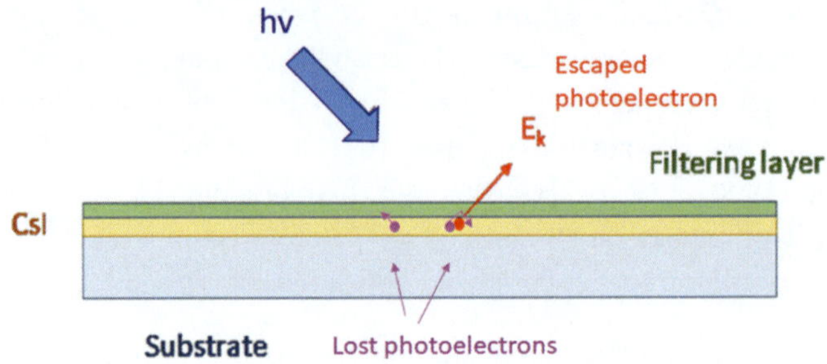

Figure 12-15. Schematic representation of a CsI photocathode coated with a thin photoelectron-filtering layer (adapted from [Abbrescia, 2020, 1–23]).

It is worth noting that there have already been attempts to coat solid photocathodes, such as SbCs, with various thin films (e.g., CsI or TMAE) [Borovik-Romanov, 1994, 269-274], [Peskov, 1994, 184-188], but for a completely different purpose, namely, to protect them from direct contact with the gas, which can degrade quantum efficiency. Although these tests revealed a suppression of photocathode efficiency at long wavelengths, this effect was not exploited for indirect light filtering. However, important conclusions can be drawn from subsequent studies conducted by several groups. One key finding is that the most suitable materials for the covering layer are alkali halides and certain photosensitive liquids, such as EF, which allow for the transport of free electrons over reasonably long distances [Charpak, 1991, 623-630].

Some of the results from these studies, aimed at tuning the effective quantum yield of CsI, are shown in Fig. 12-16, with more details available in [Breuil, 2016,1–18]. As shown, a potassium iodide layer deposited on top of a CsI photocathode provides the desired effect. In contrast, even relatively thin metallic (e.g. Al) filters suppress the quantum efficiency of the photocathode at both long and short wavelengths [Breuil, 2016,1–8].

However, in principle, it is possible to optimize such metallic layers by exploiting the island-type structure that forms at small thicknesses [Peskov, 1973,197–202].

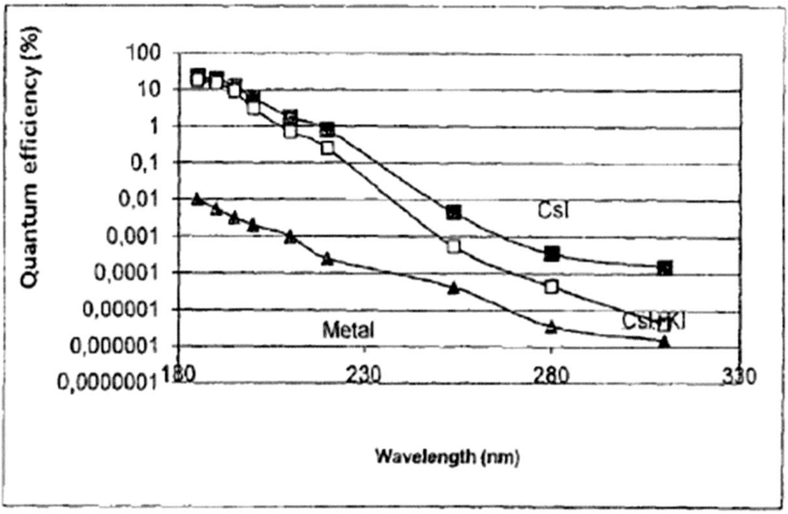

Figure 12-16. Quantum efficiency of various photocathodes: bare CsI, CsI covered with a 10 nm thick KI layer, and CsI covered with a 20 nm thick Al layer (reproduced from [Breuil, 2016, 1–18]).

12.2.5. Concluding Remarks

Despite the sad fact that Georges Charpak passed away in 2010, the collaboration he initiated remained active for several years [Peskov, 2014, 1–57], [Breuil, 2016, 1–18], [Volpe, 2018, 1–3].

New momentum was given to these studies by the EU ATTRACT project, within which significant progress was made [Abbrescia, 2020, 1–23], [Abbrescia, 2020, 1–5].

A flame detector with a CsI photocathode has been commercialized by the Finnish company Fenno-Aurum Oy Ltd. Additionally, a flame detector using EF vapours is currently in the process of commercialization in Turkey.

Work is also ongoing at CERN on the development of detectors for hazardous gases, where the photoionization method is combined with absorption measurements to improve gas classification.

CHAPTER 13

ELECTRONIC COMPONENTS AND AI: CRUCIAL ELEMENTS OF MWPCS

Abstract

This chapter explores the pivotal role of electronic components and artificial intelligence (AI) in the functionality and advancement of MWPCs. Using the ALICE experiment at CERN as a case study, we examine how sophisticated electronic systems enable precise signal detection and processing, while AI-driven techniques enhance data acquisition, filtering, and analysis. The discussion highlights how the integration of cutting-edge electronics with AI not only optimizes the performance of MWPCs but also opens new avenues for innovation in high-energy physics instrumentation. This chapter is designed for advanced readers with a background in particle physics and instrumentation, offering both a technical overview and practical insights into the complex interplay between hardware and intelligent software in the context of modern particle detectors

Even though the detector, the MWPC, remains the same, the source generating the electronic signal can vary: high-energy photons, transition radiation, Cherenkov radiation, ionizing particles, or a combination of these. Electronics designed for specific signal processing can enhance the MWPC's effectiveness in distinguishing between these different signal sources. Transition radiation produces photons with energies up to several keV, while Cherenkov photons lie in the few-eV range. Both types of photons can be converted into photoelectrons (a quantum process), whereas ionizing particles directly generate multiple secondary electrons along their paths.

With dedicated signal processing electronics and appropriate detector calibration procedures, it becomes possible to deduce photon energy or certain kinematic properties of charged particles, thereby facilitating particle identification. In the following paragraphs, three examples of specialized electronics with tailored signal processing are presented, showing how they enhance the MWPC's performance for specific measurements.

For instance, by counting transition radiation photons using simple electronic circuits, the detector can provide real-time particle identification, distinguishing, for example, protons from positrons, a capability particularly useful for fast triggering systems.

Artificial intelligence (AI), like specialized electronics, can play a pivotal role in MWPC operation and data analysis across different applications. In some complex experiments or specialized physics analyses, traditional offline data analysis based on classical software algorithms may prove inefficient, time-consuming, or prone to false positives. This is especially true in high-luminosity proton-proton (pp) or lead-lead (Pb–Pb) collisions. In pp collisions, many events occur within the same bunch crossing, with time intervals in the picosecond range, causing tracks to overlap in the detector and complicating the association of tracks to their correct interaction vertex.

In Pb–Pb collisions, each event can generate thousands of secondary particles. In such scenarios, beyond specialized electronics, the use of AI becomes an attractive option. Machine learning (ML) software, such as neural networks (a form of AI), can be trained to identify tracks belonging to the event of interest from among tens of thousands of secondary particles. This software "learns" through a training phase, allowing the ML system to outperform traditional algorithms in certain tasks. Achieving this, however, is not trivial selecting the right ML model and training approach is one of the greatest challenges before meaningful results can be obtained.

Additionally, to reduce data storage demands, ML-based event selection can be advantageous during real-time data acquisition. ML software can be deployed on Graphics Processing Units (GPUs), which allow the processing of large and complex datasets. This enables online selection, reducing the volume of data that needs to be stored. Such approaches are increasingly being adopted in new experiments at the LHC and will be essential in the upcoming High-Luminosity LHC era.

13.1 Electronic block diagram for the TRD of TS93 experiment

Chapter 7 provides a detailed description of the TS93 spectrometer. The need for low power consumption and fast, reliable electronics throughout the flight, combined with a preference for a simple readout design, made "cluster counting" the optimal solution for the TRD at that time. In this method, particle identification is based on counting the number of transition radiation photons detected per track.

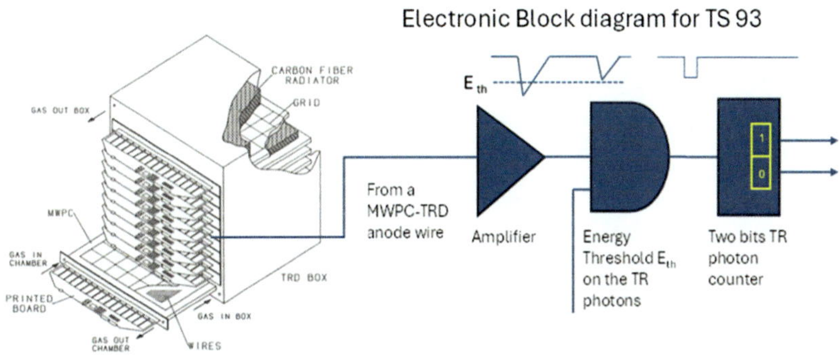

Figure 13-1. Cluster counting electronics for TR photon counter of TS93 TRD.

Fig. 13-1 shows the basic block diagram of the electronic circuit that enhances the MWPC's ability to distinguish protons from positrons and antiprotons from electrons. Each anode wire of the MWPC can detect up to four transition radiation photons. When the amplifier transmits the analogue signal from a wire to the discriminator, the discriminator outputs digital pulses whenever the input signal exceeds the preset energy threshold. A two-bit counter records the number of digital pulses during the MWPC's drift time, which lasts several hundred nanoseconds. The number of transition radiation (TR) photons detected is proportional to the particle's Lorentz factor (γ). The energy deposited by an ionization cluster is typically below the discriminator's threshold (E_{th}) for TR photon detection, except for rare high-energy delta rays that can mimic a TR photon.

Fig. 13-2 presents an example of the analog signal at the amplifier output for a radiating particle. The first negative peak corresponds to a transition radiation photon cluster, while the smaller second and third peaks represent ionization clusters.

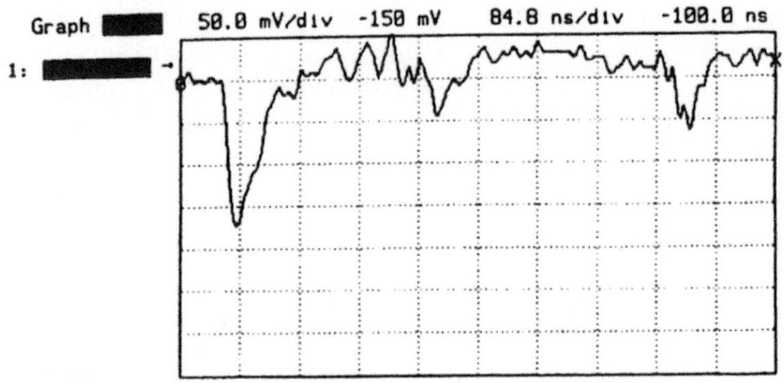

Figure 13-2. Anode wire analogue signal of a radiating particle (4 GeV/c electron). The horizontal time scale is 84.8 ns, voltage vertical scale is 50 mV/div.

To conclude, conceptually, the cluster counting electronics transforms each wire of the MWPC in a TR photon counter.

13.2. Electronic Block diagram for the ALICE HMPID at CERN

The HMPID consists of seven RICH modules, each measuring 1.3 × 1.3 m². Each RICH module contains a methane-filled (CH_4) MWPC and is equipped with six photocathodes. Each photocathode has 3,840 pads coated with a 300 nm-thick photosensitive CsI layer.

Fig. 13-3 shows one of the six photocathodes along with a single pad electronic channel. After the analogue signal from the amplifier output is digitized by the ADC, a zero-suppression circuit removes data during the cathode pad readout cycle whenever no signal is present. This approach can reduce the number of 10-bit words stored in the memory card (fourth block in Fig. 10-3) by up to three orders of magnitude, facilitating offline analysis.

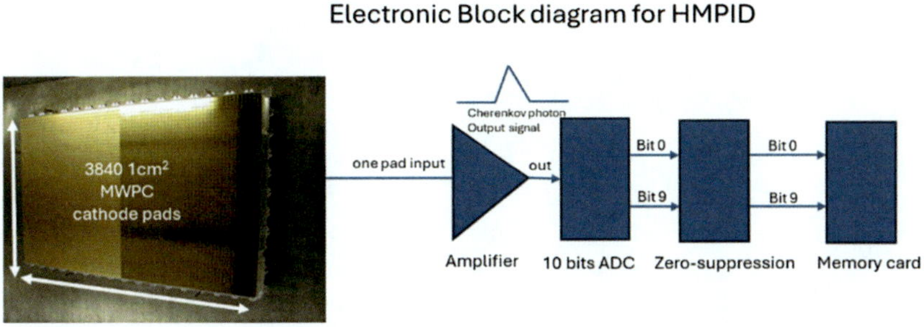

Figure 13-3 Charge analysis electronics for the MWPCs cathode pads signal of the HMPID RICH modules.

Amplifiers sensitive to individual photoelectrons (i.e., single Cherenkov photons), combined with fast and cost-effective readout electronics, enable the detector to operate in collider environments with collision rates of up to several MHz. The Cherenkov photon patterns may appear circular, elliptical, or open-conical depending on the particle's angle of incidence on the radiator. Each pad acts as an independent Cherenkov photon detector.

Fig. 7-8 previously presented an HMPID event display from a Pb–Pb collision at the CERN LHC. Thanks to the two-dimensional readout system and dedicated software, it is possible to identify Cherenkov patterns produced by primary charged particles. By combining momentum measurements from the TPC, the HMPID can subsequently determine the particle's mass and thus identify it.

Notably, in Photo 13-4 are several of G. Charpak's closest colleagues, including (from left to right) Jilbert Million, Roger Bouclier, and Jean-Claude Santiard, with whom Charpak chose to share part of his Nobel Prize. Jean-Claude Santiard, a prominent electronics engineer, developed the HMPID electronics in the late 1990s, electronics that continue to operate and collect data in ALICE as of 2025.

Figure 13-4. In the center is Nobel Prize laureate G. Charpak, accompanied by members of his former team at CERN (CERN photo.

13.3 Electronic Block diagram for the ALICE TPC (Run 2)

The operating principles of a TPC are clearly illustrated in Fig. 7-12. Electron clusters (shown as dotted vertical lines), generated along the particle's trajectory, drift toward the readout endcaps of the TPC. Upon reaching the sensing grid, consisting of MWPC anode wires, avalanche multiplication takes place, and the resulting ion motion induces a signal on the pads located directly beneath each cluster. By combining the positions of the activated pads with the drift time measured from the collision event, the TPC can precisely determine the cluster's three-dimensional position within its volume, provided the drift velocity is known. This process allows the TPC to function as a highly powerful tracking detector.

Fig. 13-5 shows a simplified block diagram of the electronic channel in the TPC MWPC. The analogue to digital converter (ADC) digitizes the positive analogue signal from the cathode pad at the amplifier output and timestamps it using the 40 MHz LHC clock, indicating the drift time required for the

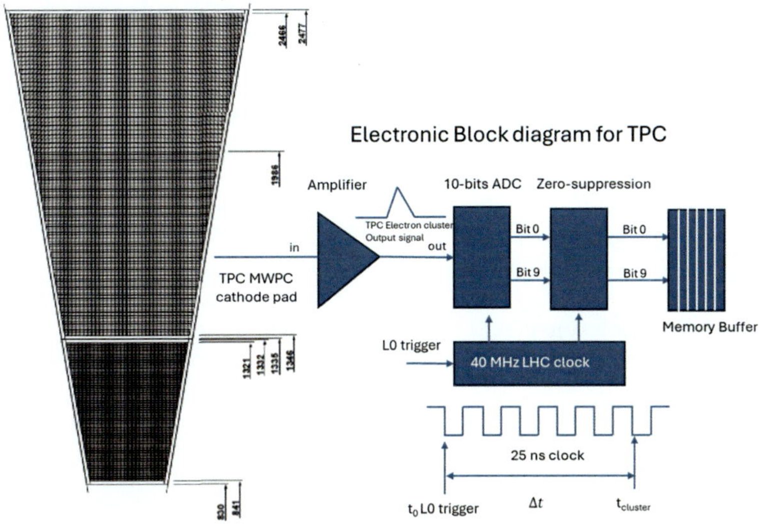

Figure 13-5. Electronic channel for signal processing of the MWPC cathode pads in the ALICE TPC endcap.

electron cluster to travel from the central gas volume of the TPC to the endcaps, where the MWPCs are located. Following a proton-proton collision, the L0 trigger (the fastest hardware trigger) initiates the digitization of the pad signal. If the signal amplitude exceeds a predefined threshold, the ADC's 10-bit word and the drift time $\Delta t = (t_{clu} - t_0)$ are stored in a memory buffer. Advanced algorithms then reconstruct the cluster's position within the TPC gas volume. If the signal does not exceed the threshold, the zero-suppression system prevents electronic noise from being recorded, conserving storage and reducing offline analysis time. Each cluster represents a segment of the particle's trajectory within the TPC. By combining these clusters, the particle's full track can be reconstructed in three dimensions. The TPC comprises a total of 570,312 electronic channels. The first column of Table 13-1 shows the distribution of pad counts at various radii across the different MWPCs in the TPC endcap.

	Pad size [mm²]	Number of rows	Number of pads
Inner chamber (84.1 < r < 132.1 cm)	4 × 7.5	64	5 732
Outer chamber (134.6 < r < 198.6 cm)	6 × 10	64	6 038
Outer chamber (198.6 < r < 246.6 cm)	6 × 15	32	4 072
TPC total		160	570 312

Table 13-1. Number of pads in the different MWPC sets. Each set (first column) corresponds to a circular ring in the endcap, positioned at a specific radius from the beam pipe.

Moreover, analysing the signal amplitude produced by each cluster provides information on the specific energy loss (dE_l/dx) along the particle's trajectory. This ionization density interpreted using the Bethe–Bloch formula, allows the determination of the product of the Lorentz kinematic variables βγ. The TPC is typically operated within a magnetic field aligned with the drift velocity. Charged particles emerging from the collision vertex follow curved paths under the Lorentz force (see Fig. 7-16). A larger curvature radius corresponds to higher particle momentum. By combining the dE_l/dx measurement with the momentum (p = mβγ), the particle's mass m can be deduced, enabling particle identification. This was the operational principle of the ALICE TPC during Run 2 [https://cds.cern.ch/record/1622286/files/ALICE-TDR-016.pdf].

Theoretically, a specialized electronic channel transforms each MWPC into a "TDC," measuring the drift time (Δt) between the particle's passage through the gas volume and the appearance of signals on the cathode pads, with timing controlled by the LHC's 40 MHz clock. Additionally, the signal amplitude digitized by the ADC enables the dE_l/dx measurement, as amplitude is directly proportional to energy loss. By integrating data from the TDC, ADC, and the track's curvature, the TPC functions as both a precise tracking system and an effective particle identification detector.

13.4. The Role of Artificial Intelligence in analysing Physics Data

Depending on its application, artificial intelligence (AI) can adopt different software architectures. In the context of high-energy physics, it is crucial to define the specific objective for which an AI model is developed. For example, in the TPC, the goal is to identify clusters corresponding to genuine particle tracks while rejecting those arising from background noise. In the HMPID, the objective is to distinguish authentic Cherenkov patterns from background-induced clusters that mimic real tracks. Such contamination, especially in Pb–Pb collisions, can degrade detector performance and limit event-by-event analysis.

It is important to note that the AI used in these applications lacks consciousness, awareness, or true understanding. Rather, it operates within an environment enriched by additional input parameters compared to classical algorithms, exploiting correlations among them. Ultimately, the expectation is that AI can achieve higher efficiency than traditional approaches.

Once the most suitable AI model is selected, it undergoes *training* on a dataset comprising high-quality tracks (including thousands of simulated events) to improve its ability to recognize genuine tracks. During this learning phase, the AI iteratively adjusts its internal parameters as it processes known particle trajectories.

After training, the AI is evaluated on an independent dataset resembling the training set but consisting of different events. This *validation phase* assesses the AI's accuracy in identifying true tracks, verifying both the effectiveness of the training process and the quality of the chosen AI model.

The third phase occurs when the AI is applied to experimental TPC data containing both real particle tracks and background clusters. Here, the AI's task is to reconstruct true trajectories while rejecting false tracks arising from background. Physicists expect that AI's background-rejection capabilities will surpass those of classical algorithms.

Classical algorithms cannot be 'trained' to discriminate genuine tracks from deceptive background patterns; they execute deterministic operations on CPUs. By contrast, a well-trained AI can leverage Graphics Processing Units (GPUs) to process '3D images' of highly complex Pb–Pb collision events (see Fig. 13-6), characterized by large numbers of secondary particles and background clusters. This ability makes AI a more powerful and efficient solution for event reconstruction than traditional methods. Neural networks have been explored to enhance clusterization in the ALICE TPC. Clusterization refers to the process of grouping localized charge deposits along a particle's path into clusters, which are then used for track reconstruction.

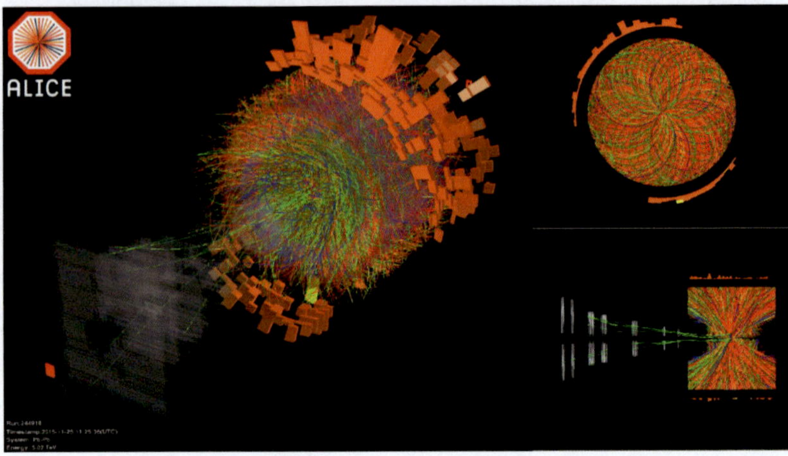

Figure 13-6. ALICE event display of detected particles created in a Pb–Pb TeV collision. Image credit: CERN.

As shown in Fig. 13-7, at a similar efficiency of ~85% for both algorithms (where the dashed line intersects the blue curve), the neural network achieves a lower fake rate (red curve) of approximately 7% (right axis), compared to ~11.5% for the classical algorithm (dotted line), a notable reduction.

In addition to lowering the fake rate, the neural network reduces the total number of clusters retained in the effective data volume by 20% by eliminating clusters not contributing to track reconstruction. This leads to improved and faster tracking performance overall.

The neural network is designed to run on GPUs during ALICE's real-time data processing at data rates up to 900 GB/s (by comparison, a typical consumer laptop or PC handles 500–1000 MB/s). It offers the potential to improve data quality while simultaneously reducing data volume.

A comprehensive presentation of this work was delivered at the Computing in High Energy Physics (CHEP) 2024 conference and is available at: https://indico.cern.ch/event/1338689/contributions/6015422/.

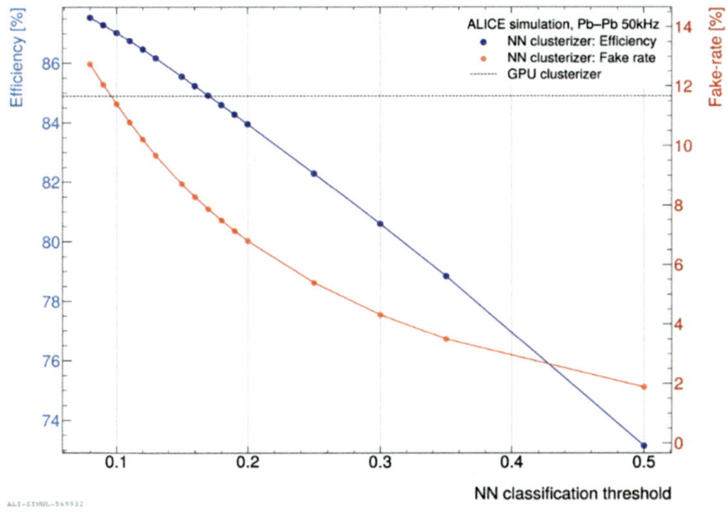

Fig. 13-7. Evaluation of the neural network clustering effectiveness in relation to the GPU cluster finder (dashed line). Explanation in the text.

The HMPID has also investigated potential improvements in Cherenkov pattern recognition using AI, comparing its performance to that of the classical algorithm known as the Hough Transform Method (HTM). Preliminary results comparing the PID performance of HTM and AI in simulated Pb–Pb events are shown in Fig. 13-8.

In Pb–Pb collisions, background clusters can mimic Cherenkov patterns from real particles, reducing the purity of the identified particle sample. In the contamination matrix, the main diagonal (dark blue boxes) represents the detection efficiency for the three particle species, pions, kaons, and protons, while the off-diagonal elements (light blue boxes) indicate contamination, i.e., misidentifications.

In these initial results, the AI (right panel) achieves higher efficiency and lower contamination than HTM (left panel). For instance, with AI, the pion identification efficiency is 96.7% with a kaon contamination of 3%, compared to HTM's 84.1% efficiency and 10.1% contamination demonstrating a more efficient and purer final sample. This is a highly encouraging outcome.

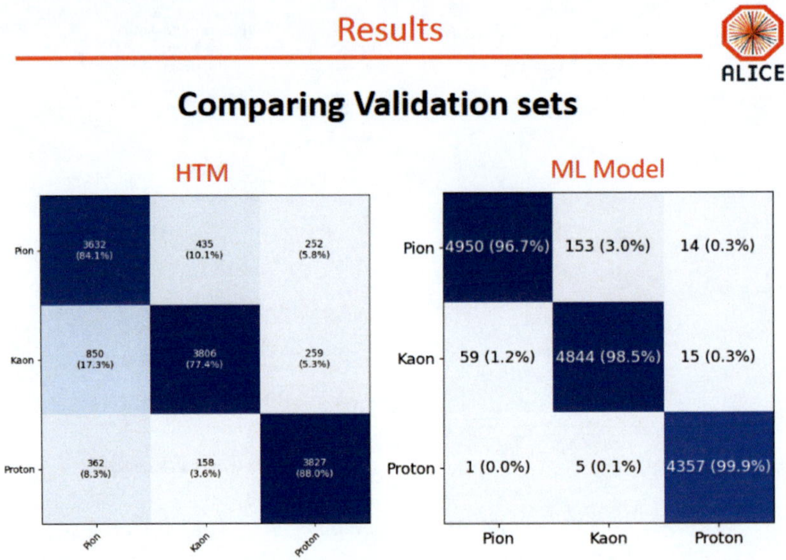

Figure 13-8. Contamination matrices. Main diagonal shows particle detection efficiencies; off-diagonal shows contamination rates.

Until now, performance evaluations have been performed via offline analysis. However, deploying AI software in real time on GPUs shows promise for reducing data volume during online data-taking, by filtering out false patterns before committing data to mass storage.

For a general overview of AI concepts and definitions, see: https://en.wikipedia.org/wiki/Intelligent_agent.

Concluding Remarks

The innovative applications of the MWPC would not have been as widely realized without the parallel progress in electronic components and computing. Similarly, the development of next-generation detectors, such as the GEM-based TPC and ALICE 3, has been intrinsically linked to advances in electronics, computer science, and artificial intelligence; without these, such rapid progress in experimental physics and related fields would not have been possible.

The integration of AI tools, as demonstrated in the ALICE TPC and HMPID, shows great promise. However, optimizing AI modules not only requires training the algorithms but also demands a learning process for physicists themselves, along with significant research and development efforts to fully harness these emerging technologies.

FINAL THOUGHTS

We have traced the evolution of gaseous detectors from their inception, highlighting the breakthroughs of the MWPC and its descendants in physics and everyday life over the past fifty years.

Many experimental physicists, engineers, and technicians played an active role in this complex endeavour, reaffirming, among other values, the fundamental importance of teamwork in the scientific enterprise.

The invention of the MWPC by Georges Charpak in 1968 revolutionized particle detection, marking a major technological and methodological leap forward. Thanks to the widespread adoption of solid-state devices (transistors, integrated circuits, et), which equipped MWPC wires with amplifiers, readout electronics, and early computers for data storage, this detector could handle significantly higher rates of particle interactions than any of its predecessors. Its ability to reconstruct particle tracks in three dimensions through fast electronics and computing marked a turning point that reshaped the field of particle detection.

The MWPC's capabilities were pivotal for high-energy physics, nuclear physics, and astrophysics, enabling the investigation of rare phenomena and short-lived particles that required large volumes of recorded events for offline analysis. The MWPC allowed immediate data collection and event evaluation. The electronic event display enabled instant visualization of particle trajectories, a significant advance over the manual photographic methods previously employed.

The modular design of MWPCs made them highly adaptable and scalable. They could be configured in various shapes and sizes to meet diverse experimental needs, from small laboratory setups to large-scale high-energy physics experiments like those at CERN. MWPCs could be seamlessly integrated with other detector systems, such as calorimeters and time-of-flight systems, or paired with radiators to function as Transition Radiation Detectors or Ring Imaging Cherenkov Detectors. This versatility enabled comprehensive experimental setups for precise particle identification and energy measurement.

The application of MWPCs led to many significant discoveries in particle physics. They played a critical role in experiments leading to the discovery of the charm quark and the identification of the W and Z bosons at CERN, contributions recognized with the 1984 Nobel Prize in Physics. MWPCs also made substantial contributions to the study of antimatter in primary cosmic rays.

The invention and use of the MWPC catalysed further advances in electronics, data acquisition systems, and computational techniques in physics. At the same time, this detector continues to benefit from the integration of artificial intelligence components, which are further enhancing its performance in analysing vast quantities of physics data. These technological developments have also influenced fields beyond particle detection, impacting medical imaging and radiation monitoring. Today, another invention by Charpak, the MICROMEGAS detector, is gaining prominence in a growing range of applications.

What does the future hold for gas detectors, and for the MWPC and MICROMEGAS specifically? These detectors now face strong competition from solid-state sensors, which have already replaced MWPCs in some areas, including tracker devices and medical imaging. The shift toward solid-state detectors is driven by advances in semiconductor technology, offering improved performance, smaller size, greater stability, and enhanced radiation resistance in various scientific and industrial applications. This transition marks a significant step forward in detector technology, enabling more precise measurements and expanding the frontiers of scientific and engineering research.

The upcoming ALICE 3 experiment, expected to begin with LHC Run 5 around 2036, exemplifies this shift. The current GEM-based TPC will be retired at the end of Run 4 and replaced entirely by solid-state detectors: a vertex detector and an outer tracker. Current prototypes demonstrate that 25-micron-thick silicon pixelated foils, incorporating front-end electronics, can be assembled into large areas spanning tens of square meters and curved into cylindrical shapes to function as time-of-flight planes or TPCs.

As we conclude this book, a question arises for the reader: has this book fulfilled its promise to spark curiosity? Has it motivated you to explore experimental physics and the development of particle detectors more deeply?
If so, we consider that a great success. If not, we still hope that learning about the history of this invention, and about Georges Charpak as both a convivial man and a brilliant scientist, as portrayed in his biography, has inspired you to join us in paying tribute to this Nobel laureate.

APPENDIX

FOR CURIOUS READERS: FURTHER EXPLORATION

Advanced Note on TPC Technology

This section diverges from the main thread of the book and is intended for advanced and curious readers. A new approach in the TPC technique, successfully implemented in the ALICE experiment, is described below: the use of GEMs as position-sensitive elements and continuous data readout.

At the beginning of the previous chapter, the crucial role of the MWPC within the ALICE TPC is presented. However, the ambitious physics program for Quark-Gluon Plasma (QGP) studies planned for LHC Run 3 (2022–2025) required significant improvements to the TPC's performance. To meet these demands, GEM detectors are replacing the aging MWPCs, and a new generation of electronics is being deployed to support continuous readout from the GEMs.

The ALICE TPC upgrade is designed to handle interaction rates of up to 50 kHz in Pb–Pb collisions during Run 3, a tenfold increase over the Run 2 rate, as detailed in [https://cds.cern.ch/record/1622286/files/ALICE-TDR-016.pdf].

With this upgrade, the MWPC has reached the end of its operational life. Looking ahead, the focus shifts to continuous (triggerless) data acquisition in the enhanced GEM-based TPC and to the development of online data analysis techniques leveraging Artificial Intelligence.

In Run 2, the ALICE TPC employed triggered readout electronics for the MWPCs for the last time. A trigger signal (L0) was generated upon the detection of secondary particles from a collision. Dedicated fast electronics further processed the L0 trigger, applying additional selection criteria to generate a Level 1 (L1) trigger signal that fired only for events relevant to the physics program. The L1 trigger then initiated the TPC readout; selected events were recorded and stored for offline analysis. This setup enabled the TPC to sustain a maximum readout rate of about 2 kHz during Pb–Pb collisions.

For Run 3 and Run 4, ALICE's physics program aims to probe the Quark–Gluon Plasma in greater detail. One objective is to study the production of heavy quarks (c and b) in particles emitted close to the beam axis (i.e., with low transverse momentum, p_t). Such measurements at low p_t are typically characterized by low signal-to-background ratios, making them unsuitable for standard low-level hardware triggering. In other words, no hardware trigger can effectively select these events.

To overcome these limitations, the following solutions have been implemented:

- Continuous readout mode for physics data, employing electronics capable of capturing 'videos' of collisions rather than 'snapshots,' introducing an entirely new data acquisition paradigm.
- Replacement of MWPCs with GEMs (mentioned in section 1.5) to reduce ion backflow into the TPC drift volume.
- Implementation of highly efficient data compression techniques to mitigate the enormous data volume generated.
- Execution of online calibration and reconstruction on a distributed computing network, aimed at reducing the amount of data ultimately stored on mass storage.

The following section provides a simplified overview of the operational principles of the GEM (which is now used for the TPC readout) and the continuous data readout. Further details on these solutions can be found in the ALICE TPC upgrade proposal [https://cds.cern.ch/record/1622286/files/ALICE-TDR-016.pdf. The principle of electron multiplication in a GEM is illustrated in Fig. 1 and 2. Electrons passing through the microscopic holes of the GEM, under the influence of a strong electric field, undergo avalanche multiplication.

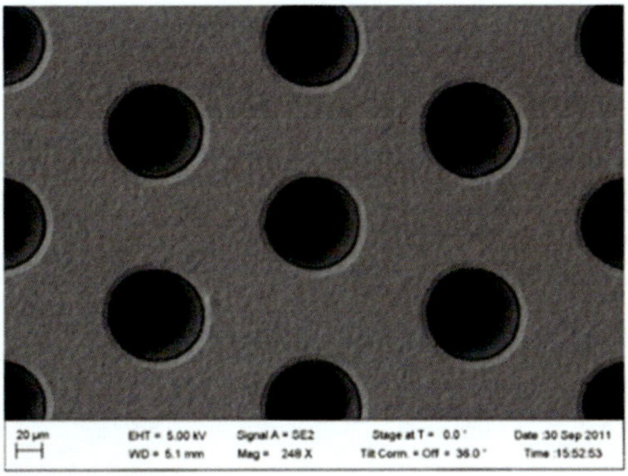

Figure 1. Electron microscope image of a standard GEM foil, with a hole pitch of 140 µm (image from CERN GDD group).

A stack of four GEMs (Fig. 3) achieves the target gas gain of 2×10^4. The pad plane located at the base of the assembled structure.

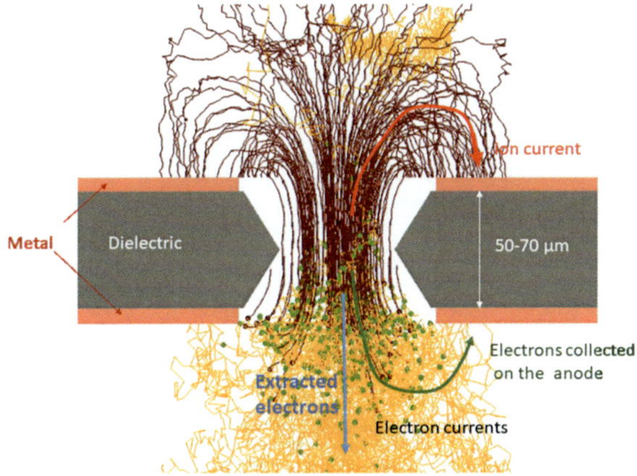

Figure 2. Garfield/Magboltz simulation of charge dynamics for two electrons entering a GEM hole. Light lines represent electron drift paths, while dark lines indicate ion drift paths. Dots mark ionization (multiplication) events. The trajectories are projected onto a cross-sectional plane. Electron and ion currents are also illustrated. Adapted from [Böhmer, 2013, 101–108].

From the pad plane, the positive electrical signal is processed by the front-end electronics card equipped with SAMPA ASICs.

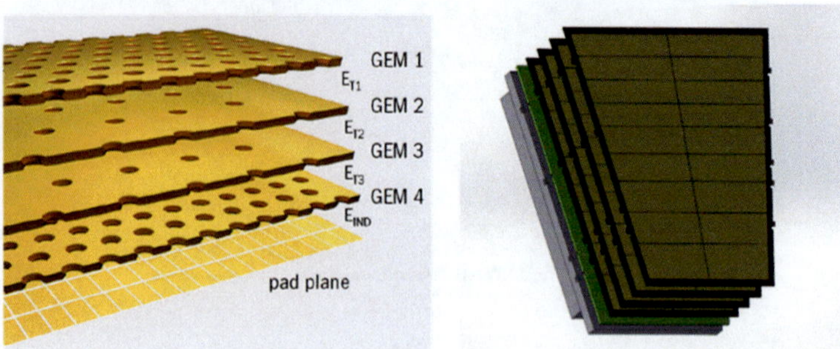

Figure 3. Artistic rendering of a stack of four Gas Electron Multipliers (GEMs). The pad plane, which forms the lowest layer of the assembled structure, is also shown. Images courtesy of CERN.

Fig. 4 presents its block diagram, highlighting the key components: the amplifier (Charge Sensing Amplifier and Shaper) and the ADC for analog-to-digital conversion.

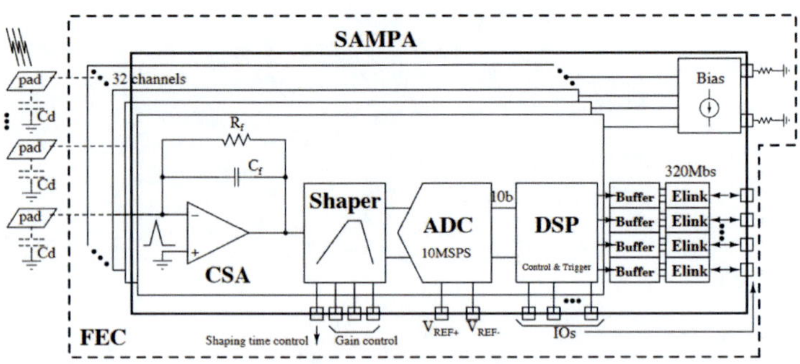

Figure 4. Schematic of the SAMPA ASICS for the GEM TPC readout, showing the main building blocks.

If the upgraded TPC (2022–2025, Run 3), equipped with GEMs and continuous readout electronics, can indeed operate at a luminosity ten times higher than in Run 2, this increase in luminosity, due to significant space-charge effects in the drift region, necessitates highly sophisticated online and offline calibration procedures. Some preliminary progress in this direction can be find in the following publications: [Adolfsson, 2021, 1-87], [Alme, 2023,1-28], [Acharya, 2024, 1-153].

Real-time adjustment of the detector response function has become crucial to effectively correct for space-charge distortions caused by ion buildup during high-luminosity LHC operations.

In Fig. 5, the complex software framework for online event reconstruction is shown, organized into two main stages. Arrays of GPUs and CPUs, both on the electronics boards and in the online computing systems, carry out various processing tasks.

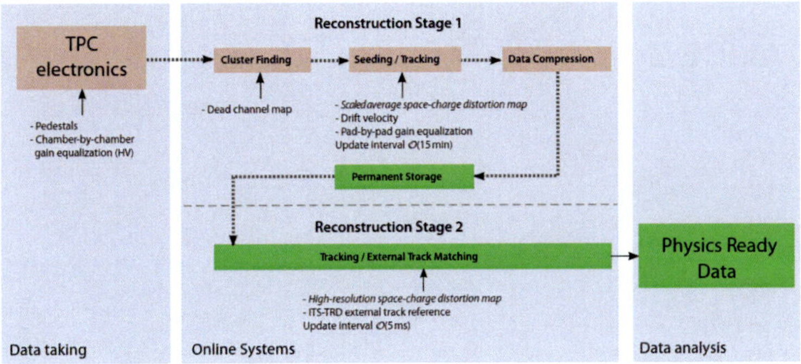

Figure 5. Schematic representation of the calibration flow during data taking and reconstruction.

The third stage of the data flow, shown in the rightmost section of the diagram, represents the offline data analysis phase, where physicists conduct their research. Access to this data and its analysis is possible via the internet, using advanced software running on a vast network of CPUs (central processing unit) distributed worldwide (the ALICE Grid). The data collected in a single year can reach 10^{12} events; without this global computing infrastructure, completing the full analysis would take tens of thousands of years.

In LHC Runs 5–6 (expected to begin around 2036), the next-generation ALICE 3 experiment (see Fig. 6) is scheduled to start operations [https://cds

.cern.ch/record/2803563/files/2211.02491.pdf]. The current GEM-based TPC will be decommissioned at the end of LHC Run 4 and fully replaced by solid-state detectors, comprising a vertex detector and an outer tracker. Prototyping efforts so far have demonstrated that silicon pixelated foils, just 25 microns thick and incorporating some front-end electronics, can be manufactured over areas of several square meters and shaped into cylindrical geometries, enabling them, for example, to function as TOF planes. The experiment will retain a cylindrical structure, approximately 5 meters in length and about 4 meters in radius, surrounding the beam pipe.

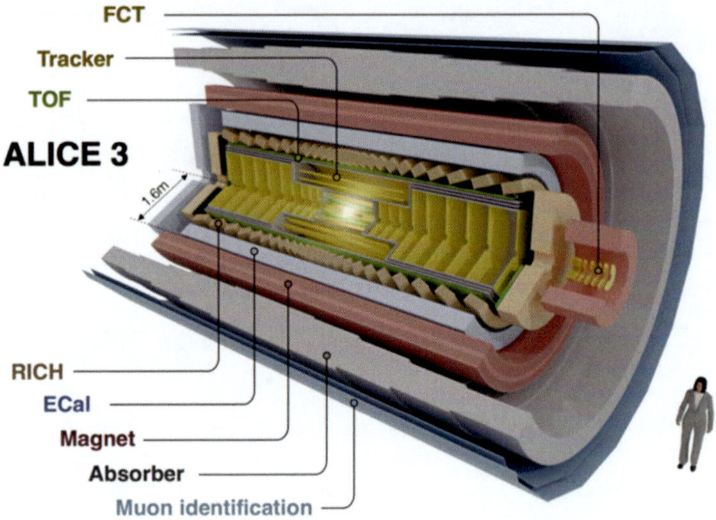

Figure 6. Longitudinal cross-section of ALICE 3. The various detector technologies are illustrated: the vertex detector, positioned closest to the beam pipe (at a radius of ~20 cm); the outer tracker, extending from ~0.2 m to ~0.8 m in radius; and the outermost muon identification system, located at a radius of ~3.5 m. FCT stands for Forward Conversion Traker. Image credit: CERN.

The time resolution of the TOF system is expected to reach approximately 20 ps (r.m.s.), enabling it to effectively replace the particle identification capabilities previously provided by the TPC. The TOF planes, designed to detect the passage of ionizing particles, will integrate one of the following sensor technologies, currently under evaluation: fully depleted CMOS sensors, Low Gain Avalanche Diodes , or Single Photon Avalanche Diodes [https://alice-collaboration.web.cern.ch/node/34993]. In this experiment, no gas detectors will be used; instead, advanced on-board artificial intelligence will play a crucial role in reducing the volume of experimental data to be stored.

BIOGRAPHY OF GEORGE CHARPAK

This biography of Georges Charpak is based on both well-known facts and private communications with Georges himself.

Georges Charpak was born on March 8, 1924, in Dąbrowica, Poland (now a region in northern Ukraine close to the border with Belarus), into a Jewish family. Some sources mention another date, August 1, 1924, as his birth was officially registered on that date.

In 1931, his family immigrated to France in search of better opportunities. His life in France became full of adventures and intriguing secrets. In a private conversation with one of us (VP), Charpak admitted that he fell in love with France at first sight and did everything possible to quickly assimilate. Their initial place of residence was Paris, where he quickly made many friends among the local boys. Georges often invited them to his home, where his mother generously fed them Ukrainian food.

Like other active French teenagers in 1937, Georges joined the "Red Falcons" movement and later the "Youth Hostels." In June 1940, the northern part of France was occupied by the Germans, dramatically affecting his family's life.

To escape the Nazi raids on Jews, they had to flee from Paris. For added safety, Georges arranged for a false identity card under the name Jacques Charpentier, which registered him in Troyes. In 1942, he lived in Montpellier in the south of France and attended preparatory classes at the high school there. His father worked as a lumberjack in the Cévennes in south-central France to remain unnoticed due to his nationality and strong accent, but the situation quickly worsened.

As the Nazi occupation of France intensified (with the Germans occupying the entire French territory in November 1942), Georges became actively involved in the Resistance movement. His participation was motivated by patriotism, a desire to fight against the oppression and brutality of the Nazi regime, and the personal stakes of being Jewish in an increasingly hostile Europe

His involvement included various clandestine activities designed to undermine the Nazi occupation and support the Allied war effort. These activities often entailed significant risks, including distributing anti-Nazi propaganda, gathering intelligence, and helping to sabotage German operations.

In 1943, Georges succeeded in passing the entrance exam and was admitted to the École des Mines in Paris. However, during the summer, he was arrested by the police, interrupting his studies. He was first interned in Montpellier prison until December when he was transferred to the Eysses detention centre. In October 1943, Eysses received most French political prisoners.

According to Georges, a completely unusual situation arose there: the prisoners managed to rally around committees from the National Front, a resistance organization, and received a softer regime of detention from the manager. They published newspapers and even managed to form an armed underground battalion.

Tragedy struck in February 1944 when a collective prisoner escape attempt failed, resulting in the deaths of some participants. As punishment, Georges Charpak was sent to Dachau concentration camp near Munich in Germany, one of the most notorious camps operated by the Nazis. The conditions were brutal, and survival was a daily struggle. Thousands of prisoners were murdered or tortured, and Nazi doctors performed inhumane medical experiments on prisoners. When Georges arrived, the camp was overcrowded, leading to disease epidemics. He remained there for a year, and according to him, his knowledge of several languages contributed to his survival.

Georges told one of us (VP) that the Nazi overseers often used him as a translator. One day, a German officer gathered all the prisoners and began gesticulating to explain that they had to work, work, and work. While translating, Georges mimicked the officer's facial expressions and gestures. Only later did he realize the risk he was taking. Fortunately, no one, including the prisoners, laughed, saving him from very serious trouble. When Georges was asked how he endured all this suffering, his answer was laconic: “I was young, physically strong, and optimistic.”

In 1945, the long-awaited day finally came: Dachau was liberated by American soldiers, who were shocked by the horrors they saw there. Georges Charpak's contributions to the resistance were recognized by the new French authorities; he received several awards, including the Croix de Guerre, and was promoted to the rank of lieutenant in the French internal troops.

After the war, Georges immersed himself in his studies, and in 1947, he graduated from the École des Mines. In his book, “Research on Particle Imaging Detectors,” Georges wrote: “In 1948, I joined the laboratory of F. Joliot-Curie at the Collège de France in Paris, being almost totally inexperienced in physics owing to serious disturbances in my life during the war.” Charpak quickly demonstrated his abilities in physics, and in 1954, he obtained his doctorate in science.

Frédéric Joliot-Curie, a famous French scientist and a prominent figure in the French Communist Party, was awarded the Nobel Prize in Chemistry in 1935, jointly with his wife Irène.

Following the war, Georges Charpak became a member of the French Communist Party, the largest party in France at that time. His alignment with the Communist Party was not unusual for intellectuals and activists in the post-war period, many of whom were drawn to the party's ideals of equality and social reform. Charpak's involvement in the Communist Party was driven by his desire to contribute to the reconstruction of a more just and equitable society in the aftermath of the war.

According to Georges, he left the organization in 1956 as a protest against the Soviet intervention in Hungary. However, he remained politically active throughout his life

One of the CERN scientists, Lucien Montanet, told us that in the late fifties, he once met Georges at the entrance of the Collège de France, surrounded by technicians whom he was encouraging to go on strike.

Following the recommendation of Joliot-Curie in 1959, Georges entered CERN and became a member of the "g-2" experiment, proposed earlier by Leon Lederman. In the 1950s, the muon was still a mystery. Physicists were not sure whether it was a much heavier electron or whether it belonged to a new family of particles. The "g-2" experiment aimed to measure one of the properties of this strange electron – its magnetic moment

Photograph 1. Team members of the "g-2" experiment at the Synchrocyclotron in 1961. Francis Farley stands next to (left to right) Johannes Cornelius Sens, Georges Charpak, Théo Müller and Antonino Zichichi, sitting on the experiment's 6-metre-long magnet. CERN image.

While working on this experiment, Charpak became close friends with Lederman and team member Richard Garwin; this friendship lasted for life.

Georges was a member of the "g-2" experiment for several years; however, his interest gradually shifted towards instrumentation. In 1965, a Georgian physicist named Chikovani came to CERN as a visiting scientist and introduced Georges to new electronic tracking detectors: spark chambers and streamer chambers.

Charpak's acquaintance with Chikovani was a turning point in his career. At that time, CERN physicists enjoyed a great deal of creative freedom, which helped to reveal Charpak's talent as an inventor. Working with spark chambers and studying their shortcomings, he began thinking about creating a more advanced detector that would not require pulse high voltage and could measure the energy losses of particles.

His experience working with single-wire proportional counters in Paris proved invaluable. This led to the gradual development of the idea of a multi-wire proportional chamber.

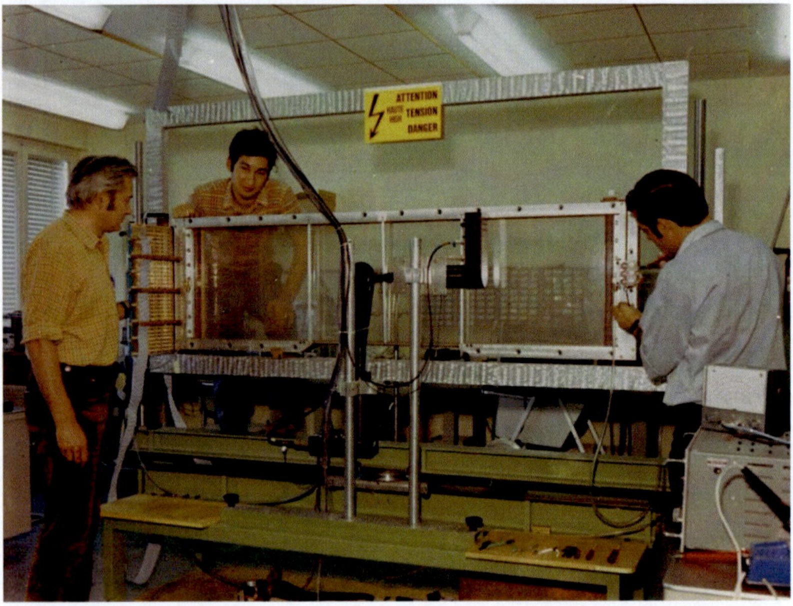

Photograph 2. Left to right: G. Charpak, F. Sauli and J.C. Santiard testing a large MWPC at CERN in 1970 (from https://cerncourier.com/a/georges-charpak-a-true-man-of-science/).

The first prototype of Charpak's multi-filament wire chamber was only 10x10 cm² in size, but it successfully demonstrated proof of concept. It was the fastest imaging detector of its time, allowing continuous recording of large flows of ionizing radiation, such as elementary particles, x-rays, and gamma photons. This invention immediately resonated within the high-energy physics community at CERN and beyond. Georges formed a group of assistants and followers who helped bring his ideas to life, solidifying his reputation as a prolific generator of ideas.

Soon, many other independent groups joined the development efforts, often competing with Charpak. As a result, MWPCs quickly became essential components in most high-energy physics experiments. The MWPC technology facilitated several fundamental discoveries, including the charm quark, W and Z bosons, and the top quark.

From the beginning of 1980 until 1991, Charpak's group focused on developing photosensitive MWPCs for ring imaging detectors and calorimeters. These studies culminated in the creation of large RICH detectors, such as the DERHI RICH detector filled with TMAE and the ALICE RICH detector combined with a CsI photocathode.

In 1992, Charpak was awarded the Nobel Prize in Physics "for his invention and development of particle detectors, in particular, the multiwire proportional chamber."

Photograph 3. The Nobel Prize ceremony: King Carl XVI Gustaf of Sweden presents a Nobel Prize medal to Georges Charpak. Image credit: CERN.

Photograph 4. Three Nobel Prize winners at CERN, Calo Rubbia (left), Samuel Ting (centre), and Georges Charpak (right) seen at the party to celebrate Georges Charpak's Nobel Prize in Physics. Image CERN.

Following this recognition, he became a public figure and dedicated some of his time to various public events. Nonetheless, he continued his scientific activities. He led developments at the company Biospace and organized an instrumentation group in Saclay, France. In 1994, he made another significant invention: a micropattern detector for high-flux ionizing particles, later named MICROMEGAS. This detector has found numerous applications both in high-energy physics and beyond.

In the last years of his life, Charpak became interested in the Hands-On Physics project. Originally initiated by Leon Lederman in the USA, the project aimed to attract talented schoolchildren to physics.

Georges Charpak's involvement in the "Hands-on Physics" program stemmed from his firm belief that traditional methods of teaching science were insufficient to inspire and adequately educate young students. He advocated for an approach where children could learn science by doing, engaging in experiments, asking questions, and discovering principles through practical experience rather than rote memorization. Implementing this project required developing a specialized school program and courses for retraining teachers.

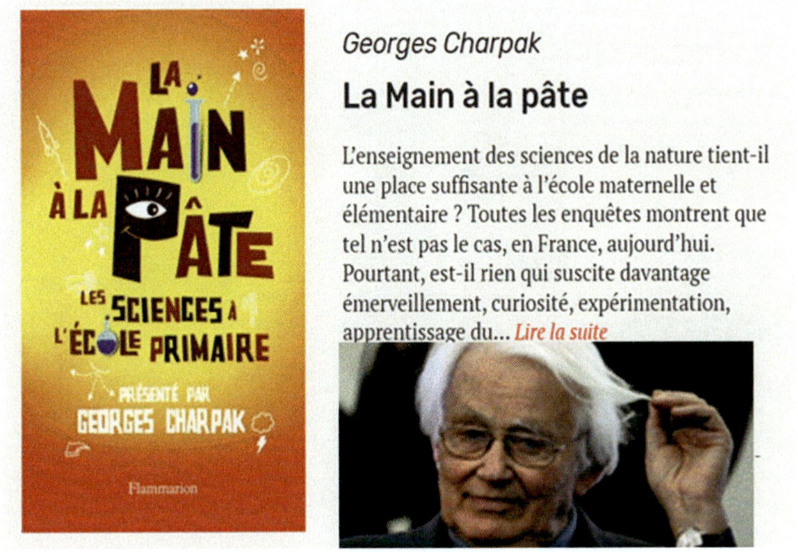

Photograph 5. Georges Charpak (CERN image) and his book "La Main a la Pate."

In the last ten years of his life, Charpak wrote about a dozen books on various topics, ranging from autobiographical works to discussions on the nuclear threat, disarmament, and hands-on physics.

Charpak passed away on September 29, 2010, in Paris at the peak of his fame.

Photograph 6. Opening of Charpak Street at CERN. From left to right: Sigurd Lettow, Dominique Charpak and the mayor of Saint-Genis-Pouilly. Image credit: CERN.

Some Fun Stories About Charpak

A Letter to Georges

One of the unique features of the Moscow Institute of Physics and Technology (MFTI), where I enrolled in 1965, was its strong emphasis on hands-on research. Beginning in the second year, students were required to spend an increasing amount of time working in some of the finest laboratories in Moscow and the surrounding area.

My first practical experience took place at Korolev's Space Center, a hub of Soviet space exploration. Later, I was fortunate to be accepted into one of the most prestigious scientific institutions, the Institute for Physical Problems of the Russian Academy of Sciences in Moscow. Along with a select group of students, I had the honour of joining the Physics Laboratory there, led by P.L. Kapitza, as a young researcher. Kapitza needs little introduction, not only was he one of the most eminent Soviet physicists and a Nobel Prize laureate, but he was also revered for his integrity, independence, and moral courage, especially during the dark days of the Stalinist regime.

When I joined P.L. Kapitza's laboratory, he was focused on studying stationary, high-power, high-frequency discharges in hydrogen (H_2) and deuterium (D_2). Diagnosing the plasma produced by these highly unusual discharges was an exceptionally demanding task. One of the main challenges was the intense emission in the visible spectrum, which masked the short-term instabilities developing within the plasma column.

After numerous attempts, we discovered that the most effective way to visualize these instabilities was through the use of multi-wire proportional counters (MWPCs), a technique originally invented by Georges Charpak. We significantly modified Charpak's original MWPC design to suit our experimental needs.

One of our key innovations was the development of a photosensitive MWPC capable of detecting individual photoelectrons generated by ultraviolet photon absorption in a specially formulated low-ionization gas mixture. Another major advancement was the creation of an MWPC optimized for the detection of soft X-ray photons, capable of analyzing their spectra with exceptionally high energy resolution.

After several years of working in P.L. Kapitza's laboratory, I was promoted to group leader for plasma diagnostics across the visible, ultraviolet, and X-ray regions of the spectrum. Those were the happiest days of my life, the work was deeply engaging, the conditions were excellent, and I had the privilege of interacting with some of the greatest scientific minds of the era, including P.L. Kapitza, E. Lifshitz, L. Pitaevskii, I. Zeldovich, A. Borovik–Romanov, Kadomtsev, and many others.

At the time, I never imagined I would one day work abroad. Although I had received solid training in English and German at the Institute, I rarely had the opportunity to use either language. As a result, I quickly lost fluency, retaining just enough proficiency to read scientific literature.

The results we obtained with various modified MWPCs were outstanding: we succeeded in visualizing hidden plasma instabilities and even detecting ultra-weak X-ray radiation produced during the development of certain types of instabilities.

One day, if I remember correctly, it was in 1981, I decided to write a letter to Georges Charpak, describing our success in applying his MWPC technology to plasma diagnostics. It was a completely emotional gesture. I had no real expectation of receiving a reply.

First, not all letters were allowed to leave the USSR. And second, Georges was already so famous at the time. Why would he respond to an unknown researcher from Moscow?

I was astonished when, two months later, I received a reply! The secretary handed me an opened envelope from CERN, foreign mail was always inspected, and an authorized person had already checked its contents. Inside was a short-handwritten letter from Georges himself. He wrote that he was very interested in learning more about our work and, for that reason, invited me to participate in the Wire Chamber Conference in Vienna, scheduled for February 1982.

Traveling abroad for a conference? At the time, that seemed absolutely impossible. I was simply content to have received a reply, and even more so that our work had caught the attention of such a great man.

But a few months later, another letter arrived, also in an opened envelope. This time, it was from the organizing committee of the Vienna Wire Chamber Conference, with a formal invitation and an offer to fully cover the costs of the trip.

I didn't take any action, but somehow, word reached P.L. Kapitza. Soon after, I was summoned to his office.

"I'm glad your work is being recognized," Kapitza said. "We'll try to get you to the conference, but as you know, it won't be easy. Are you a member of the Communist Party?"

"No," I replied.

"That will complicate things… but we'll try," he said. "However, before you go, you must patent your photodetector."

Following his advice, I submitted a patent application. Unfortunately, the patent officer at our institution was so slow and inefficient that he never even filed it with the USSR's patent agency.

The First Meeting with Gorges at the Vienna Wire Chamber Conference

Let me skip over the difficulties I faced in obtaining permission from the Soviet authorities to attend the Wire Chamber Conference, despite the full support of P.L. Kapitza, and get straight to the heart of the story: my first meeting with Georges. It happened in February 1982, at the registration desk of the Vienna Technical University.

I approached the desk and asked the young woman registering participants, "How can I meet Prof. Charpak? I'd like to thank him for my invitation."

"There he is," she replied, pointing to a tall, broad-shouldered man. He looked like Jean Moreau, familiar to me from French films. The man was in the middle of a lively discussion with the conference organizers.

I moved closer, waiting for a pause in the conversation so I could introduce myself. But instead of winding down, the discussion grew more intense. Despite my limited English, I could understand that he was trying to secure some financial support from the organizers for his collaborator, D. Anderson

"He made a breakthrough, his talks will be the best at the conference, and you don't want to cover his expenses?" Georges argued with the group around him.

"We don't have any extra funds; you should have requested this in advance," one of them replied.

Georges grew visibly angry. "Fine, I'll pay him from my own budget, then."

Clearly, it wasn't the right moment to introduce myself, so I decided to wait for a better opportunity. But just then, the smiling girl, perhaps trying to defuse the tension, suddenly said, "Professor Charpak, this is Dr. Peskov from Moscow."

Georges turned to me with a stern expression. "I'm very busy today. Let's talk tomorrow," he said, and with that, he left the hall with D. Anderson.

I was left standing there alone and disappointed.

The next day, the conference began. Georges was seated in the front row, occasionally making remarks to the speakers. In contrast, I sat silently in the back row, feeling uncomfortable and out of place. I didn't know anyone and felt like a stranger at the conference.

At lunchtime, the same friendly girl who had helped me before approached with a warm smile. "Professor Charpak invites you to lunch," she said. "Please wait near the desk."

Soon, Georges appeared. "I'd like to discuss your work with you," he said. "Let's eat together."

Photograph 7. G. Charpak and one of the authors of this book (V.P.) at the Vienna Wire Chamber Conference, 1982. ***(****Source:* CERN Courier, *Volume 23, No. 4, May 1983, p. 138).*

We found a small, quiet restaurant not far from the Technical University. Once we were seated, Georges began, "I'm very interested in your work and in the applications of the MWPC in plasma studies. But you're not the first to develop a photosensitive MWPC. Are you familiar with this work?"

He handed me a paper by authors I didn't recognize, Ypsilantis and Séguinot.

"No, I've never read this," I replied.

"They developed a similar photosensitive MWPC about a year before you," Georges continued, "but for a different application, ring imaging."

I handed him my paper, originally written in Russian and translated into English. He flipped through it quickly.

"When did you submit this for publication?" he asked.

"About a month after Séguinot," I replied.

"That's interesting," Georges said. "So your group developed almost the same detector, but independently!

George's expression shifted from official to friendly. "Oh," he continued, "and you use another photosensitive gas." His face grew even warmer. Then, suddenly, he asked, "What do Soviet physicists think about Sakharov? Do they protest his exile in Gorky? I heard there was even an attempt to expel him from the Academy of Sciences."

It was a provocative question. During Brezhnev's time, engaging in such discussions was risky. George looked at me with piercing eyes.

After a brief pause, I replied, "Yes, I heard about this. Academician Scriabin gathered the top leaders of the Academy of Sciences and announced that there was a recommendation from the Politburo to expel Sakharov. But then he added that there was no precedent for such an action in the Academy's history, so he didn't know how to proceed. At that moment, P. L. Kapitza, the director of our institute, who was among those present, spoke up. 'No, there was a precedent!' he said. 'Einstein was expelled from the German Academy of Sciences.'"

His remark sent a shock through the room, and the discussion quickly shifted to other matters. Sakharov was saved.

"Saved?" Georges asked ironically. "But he's in Gorki."

"But even that was a heroic act," I replied, becoming increasingly emotional despite my wish to stay out of such a dangerous discussion. "Do you know that Kapitza saved several lives during Stalin's terror? Landau, Fock, they were released from prison thanks to letters Kapitza sent directly to Stalin. That was real heroism! One night, Kapitza was taken to the KGB, where he was put under immense pressure to withdraw his request for Landau's release. But he stood firm, and after six months, Landau was freed," I explained. (unfortunately, at that time, I had no idea that in 1980, P. L. Kapitza had written a letter to Y. V. Andropov, the head of the USSR KGB, requesting the release of A. D. Sakharov).

Georges looked at me with his blue eyes, and I began to feel that a quiet, mutual sympathy had begun to grow between us.

"What are you doing tomorrow evening?" he asked.

"I'm planning to explore Vienna," I replied.

"Good," said Georges. "If I understood correctly, this is your first trip abroad?"

"Yes," I answered. "And thank you, Georges."

We left the restaurant almost like friends.

The next day, the organizers of the Vienna conference gave me some money. I was staying in a small, pleasant hotel near the famous Vienna Opera, and the temptation was too great to resist, I decided to try and get a ticket for the evening ballet.

There were no tickets available, but I was incredibly lucky: someone whose companion couldn't attend sold me a spare ticket. It was very expensive, and I ended up spending all the money I had received that morning from the committee.

The Vienna ballet deserves a description of its own, it truly left me in awe. I was hungry and, with some jealousy, watched people enjoying food and champagne in the Opera cafeteria. During the intermission, I unexpectedly ran into Georges. He was well dressed and surrounded by six or seven companions. Our encounter at the ballet seemed to surprise him as much as it did me.

He introduced me to those around him.

"They're members of my group," he said.

For the first time, I met Breskin, Majewski, and several of Georges' collaborators. That evening, a small dream of mine came true, I was offered a glass of champagne.

"How did you manage to get a ticket?" Georges asked. "We requested ours two months before the conference."

"I bought mine from a private person," I replied.

"I imagine it cost a fortune?"

"Yes," I admitted. "I spent all the money I received from the conference organizers."

"And how are you going to survive the rest of your time in Vienna?" he asked.

"I'll eat what my wife packed for me," I said.

"What exactly?"

"I have some canned food and a cake I brought from Moscow."

"Aha!" Georges said, looking at me with a curious expression.

When the ballet ended, Majewski turned to Georges and said, "Why don't you invite Vladimir to CERN?"

"Good idea!" Georges replied. "But first, I want to visit Moscow, see his institute, and be introduced to Kapitza!"

When I arrived at the conference the next morning, the same girl from the organizing committee was already waiting for me.

"Professor Charpak told us you spent all your money on last night's ballet," she said with a charming smile. "The organizing committee has decided to compensate you, you'll receive some extra funds from us at lunchtime."

George's Trip to Moscow in 1984

When I returned to Moscow, I reported to P.L. Kapitza about my meetings with Georges and his desire to visit our institute. "Of course, we can invite him," Kapitza replied. "That will be a much easier task than getting permission for you to attend the conference."

Seven months later, I went to the Moscow airport to meet Georges, who had arrived thanks to Kapitza's efforts. As we left the airport, we saw a huge line for taxis. I approached a policeman directing traffic and said, "I'm accompanying a famous French scientist. Would you please make an exception and arrange a taxi to Moscow for us?"

The policeman didn't say a word but immediately stopped the first empty taxi that approached. Georges was impressed.

"How did you manage to bypass the line?" he asked.

"You're so famous in Russia that everyone wants to please you," I replied.

"You know, Vladimir," Georges said, "famous people can't plan their own time. Everything is arranged for them by assistants. So, what's our agenda for the next few days?"

"You're right, it's all pre-planned," I said. "Tomorrow, Monday, you'll visit our lab. On Tuesday, you'll meet with P.L. Kapitza. On Wednesday, as you requested, I've arranged your seminar at ITEP. Thursday, we'll visit the Space Institute, and Friday is free, we'll take a walk around the city. In the evening, you're invited to visit Kapitza's dacha."

"Tomorrow morning, Kapitza's son Sergey and I will pick you up from the hotel."

In the USSR, Sergey Kapitza was a well-known public figure, as he hosted a popular science television show. The next morning, when we entered the hotel together, the staff immediately looked flustered and excited. When Georges appeared and I introduced him to Sergey, all eyes turned toward us, accompanied by smiles and warm greetings. Georges probably realized at that moment that I hadn't been joking in the taxi, he really *was* well known in Russia, and his visit was clearly anticipated.

Buoyed by the friendly attention, we climbed into Sergey's old car and set off toward our institute. But just 50 meters before reaching the gate, the car suddenly stopped and wouldn't move any further. Sergey popped open the hood and quickly diagnosed the problem.

"The motor belt is broken. I'll steer, and you push the car," he said.

"Why don't you call a car service?" Georges asked.

"That doesn't exist in our country," Sergey replied matter-of-factly.

Laughing, Georges and I got behind the car and started pushing.

Near the entrance of the institute, a few girls were waiting, as they did almost every day, hoping to catch a glimpse of Sergey and get his autograph. As soon as they spotted him, they ran toward us. Without hesitation, Sergey jumped out of the car and slipped through a small side gate into the institute, leaving Georges and me alone with the broken-down vehicle in the middle of the road.

"What just happened?" Georges asked, puzzled.

I explained the situation to him, and he burst out laughing. In the meantime, we had lost control of the car. Even though the brake was on, it started rolling backward into traffic. It took a huge effort from both of us to stop it.

From a safe distance inside the institute grounds, we saw Sergey waving at us, signaling to abandon the car and come inside.

"Tomorrow, P.L. Kapitza will show you his gigantic plasma installations," I said as Georges entered my lab. "Today, I'll show you my small MWPCs, which we use to visualize plasma instabilities."

Georges picked up each MWPC and examined them carefully, one by one.

"Can you demonstrate that they're sensitive to single photoelectrons?" he finally asked.

I had a special ultraviolet source that emitted single photons, so we set up the experiment. After two hours of meticulous work, Georges was convinced that our detectors functioned properly.

"How do you measure their efficiency?" he asked after we took a short break for lunch.

To answer that, we brought in a reference detector, a cooled photomultiplier with a known quantum efficiency. After another three hours of detailed measurements, Georges was fully satisfied.

"I've never worked so hard in my life as I did today," he joked as we finally arrived at his hotel late in the evening.

When we arrived at the institute the next morning, we were met with very sad news: just an hour earlier, P.L. Kapitza had been taken to the hospital by emergency medical services. His son, Sergey, had gone there to speak with the doctors.

Without Kapitza, I couldn't take Georges into the plasma lab. The most I could do was briefly show him the massive high-frequency generators and plasma chambers through a slightly opened door, closely guarded by a security officer.

We then went to I. Zeldovich's office. Zeldovich, the head of the Theoretical Division at our institute, was well acquainted with the plasma research being conducted in Kapitza's lab. For about an hour, he and Georges discussed a range of scientific topics: plasma physics, fusion, cosmology, high-energy physics, and more.

Once a sense of trust had been established between them, Georges asked Zeldovich what he thought about the situation involving Sakharov and Orlov. Zeldovich had once worked closely with Sakharov on atomic weapons development and was well aware of his circumstances. Despite my presence, which was unusual, as such sensitive topics were rarely discussed in front of witnesses in the USSR, Zeldovich gave Georges a brief but very clear explanation. Like Kapitza, he was an honest and courageous man.

Georges' face grew warm with emotion, and he embraced Zeldovich. "I am with you!" he said.

I was quite nervous about the situation. Discussing such matters in Vienna was one thing, doing so in the USSR was quite another. Who knew? Perhaps someone had overheard the conversation.

The next day, I took Georges to ITEP, where L.B. Okun had arranged a seminar for him. The auditorium was packed, some people were even sitting on the floor. Georges gave a lengthy talk, presenting black-and-white transparencies in a steady, almost monotonous tone, detailing his work and the various applications of MWPCs in high-energy physics experiments. The audience listened intently, captivated by his expertise.

Then, suddenly, he projected a full-color slide showing two elephants mating. He quickly removed the image and, with a mock-startled expression, said,

"Sorry, wrong presentation."

The room erupted in laughter. It was a bold joke, especially considering that, in Soviet times, showing such images was strictly forbidden. Georges was well aware of this. But it was in his nature to challenge conventions and tease boundaries. Jokes like this created an unspoken connection between him and his audience, a kind of invisible bond of shared amusement and subtle defiance.

In the evening, Okun invited Georges and me for dinner. On the table, he had set out several glasses of red wine.

"Georges," he said, "these are Georgian wines and one French wine. Can you tell them apart?"

Georges carefully tasted each glass and replied, "There's no French wine here, but the Georgian ones aren't bad at all."

"Correct," Okun said with a smile. "I got these from my Georgian friends. They're all good and carefully selected."

Afterward, Georges and Okun began discussing Orlov's case.

"I'm planning to tease the Soviet authorities," Georges said. "I'm working on a major review paper about MWPCs, and I intend to dedicate it to Orlov. I don't think the KGB reads *Nuclear Instruments*, so the journal will make it into the USSR, and all the physicists will see the dedication, a subtle challenge to the system!"

The conversation then shifted, and Georges, along with some of Okun's invited guests, began discussing another sensitive topic: the situation of the Jewish population in the USSR. After a long and thoughtful exchange, Georges concluded with a smile:

"This topic has such a long and complicated history, people can eventually develop paranoia around it. Have you heard the story? The United Nations once declared a 'Year of the Elephant' and asked all nations to submit a report on the role of elephants in their societies. The Soviet Union submitted a paper titled *The Role of Elephants in the Soviet People's Economy*, the Germans submitted *Elephants and the Demilitarization of the Bundeswehr*, the French offered *Elephants and Sexual Fantasies*, and Israel submitted *Elephants and the Jewish Question*."

Everyone burst into laughter, and by the end of the evening, all the Georgian wine had been drunk in a warm and spirited atmosphere.

On Thursday, we spent the entire day at the Space Research Institute of the USSR Academy of Sciences, where Georges was hosted by Rashid Sunyaev. Sunyaev had also arranged a seminar for Georges. Unlike the ITEP seminar, this time Georges needed an interpreter, as not everyone in the audience understood English. The interpreter was a young blonde woman.

Georges began his talk with a thoughtful introduction: "Your institution specializes in space and cosmology. Throughout human history, people have looked up at the sky, seeking to understand its origin and that of the universe." He then displayed a cartoon: a man gazing at a star-filled sky. At first glance, it seemed innocent, but upon closer inspection, one could discern that the man was actually looking at a nude woman.

A ripple of laughter spread through the auditorium. The young interpreter blushed, but Georges, seemingly unfazed, kept the cartoon on the screen for an extended period, continuing his presentation. It was a bold move, especially considering the conservative norms of Soviet-era public discourse. Yet, this was characteristic of Georges, challenging conventions and forging a unique connection with his audience through subtle humour.

On Thursday, we spent the entire day at the Space Research Institute of the USSR Academy of Sciences, where Georges was hosted by Rashid Sunyaev. Sunyaev had also arranged a seminar for Georges. Unlike the ITEP seminar, this time Georges needed an interpreter, as not everyone in the audience understood English. The interpreter was a young blonde woman.

Georges began his talk with a thoughtful introduction: "Your institution specializes in space and cosmology. Throughout human history, people have looked up at the sky, seeking to understand its origin and that of the universe." He then displayed a cartoon: a man gazing at a star-filled sky. At first glance, it seemed innocent, but upon closer inspection, one could discern that the man was actually looking at a nude woman.

A ripple of laughter spread through the auditorium. The young interpreter blushed, but Georges, seemingly unfazed, kept the cartoon on the screen for an extended period, continuing his presentation. It was a bold move, especially considering the conservative norms of Soviet-era public discourse. Yet, this was characteristic of Georges, challenging conventions and forging a unique connection with his audience through subtle humor.

On Friday evening, Georges experienced Russian hospitality as we visited the Kapitza family's dacha (summer house). Although P.L. Kapitza was still hospitalized, his son Sergey and other family members warmly welcomed us. After dinner, Sergey gave Georges a tour of his father's private laboratory.

"After World War II, Stalin asked my father to work on the atomic bomb project," Sergey explained. "He refused, not because he objected to the bomb itself, but because he couldn't tolerate Lavrentiy Beria, the head of the Soviet secret police (NKVD), who was also in charge of the project. As a result, my father was removed from his position as director of the institute and placed under house arrest at our country dacha. Determined to continue his experimental work, he built a small laboratory with the help of friends and his own funds. Here, he began developing high-frequency generators."

Sergey showed Georges some of these homemade generators. "Later, my father started investigating plasma produced in high-frequency cavities," he continued. "I've never heard about this," Georges replied. "I'm truly impressed. Your father is a remarkable man, I'd be honored to shake his hand someday!"

My First Days in George's Lab as VP (1986)

Unfortunately, George's dream of meeting P.L. Kapitza never came true. After a period of hospitalization, Kapitza's health declined significantly, and he passed away in 1984. Plasma research at our institution was subsequently discontinued, and I could no longer pursue the work I had enjoyed so deeply.

Around that time, Georges sent me a letter inviting me to join his lab at CERN for a year. I saw it as a valuable opportunity to begin a new chapter in my life, and I decided to take it.

Going abroad for work, it was an incredibly difficult thing to arrange at that point in history. It took Georges and me two years to secure permission from the Soviet authorities for me to travel to CERN. Even then, to ensure I remained under their control, they refused to let my family accompany me. My wife, son, and daughter had to stay behind in Moscow, effectively held as "hostages."

My first day in Georges's lab was extremely challenging. Everything was so different from what I knew, the people, the customs, even the lab equipment. I experienced a profound culture shock.

Georges understood very well how difficult it was for me to adapt, and he was always kind and empathetic. One memorable example stands out.

One day, I called my family in Moscow from CERN. Wanting a bit of privacy, I closed the door to my office. It was an emotional call, I spoke with my wife, then my son, and finally my little daughter. They were all struggling without me. Tears welled up in my eyes.

Suddenly, the door opened and Georges walked in. He saw me on the phone, tears on my face, and quietly stepped out without saying a word. An hour later, he returned, looked at me, and said gently, “Today, you and I are going to a very nice French restaurant. I’ve already made a reservation.”

“But Georges,” I replied, “I don’t have a French visa, only a Swiss one!”

“I’ll drive you in my car,” he said with a reassuring smile. “It has diplomatic plates. Nobody can stop us.”

And so we went. The restaurant was indeed beautiful. As we sat down, Georges said, “We’ll drink some good French wine, and you’ll see, it wasn’t a mistake to come to Geneva and work with my group, even if you miss your family sometimes.” Then he turned his attention to the long list of wines the waiter had just handed him.

After some careful deliberation, Georges finally selected a bottle of wine. The waiter soon returned with it, opened it, and offered Georges a taste.

Perhaps wanting to lift my spirits and distract me from thoughts of my family, Georges leaned over and said with a playful smile, “You know, Vladimir, in France you can ask for a different bottle if you don’t like the first one, at no extra cost. To cheer you up, let’s put on a little show!”

He proceeded to smell the wine, take a sip, then smell and taste again, clearly playing up the role. Finally, he said something to the waiter in French, then turned to me and translated: “The wine is not as good as it should be.”

The waiter immediately began to argue, clearly insisting the wine was fine. “We’ll call his boss,” Georges said confidently, and the conversation continued in French.

A minute later, the manager arrived, tasted the wine himself, and said something in agreement. Georges looked pleased. “The boss said the wine is actually good, which is true, but he’ll bring us another bottle anyway.”

This little performance truly lifted my mood. The rest of the evening was filled with laughter, excellent food, and, of course, good French wine. Yes, perhaps it wasn't a mistake to come to Geneva and join Georges's group after all.

Photograph 8. The Charpak group in 1986. Image credit: CERN.

Georges's Dinner at My Mother's Apartment (1986)

George's last trip to Moscow took place when I was already working in his lab. This time, he had received an invitation from E. Velikhov, a vice president of the USSR Academy of Sciences, an honor he was very proud of. At every opportunity, whether appropriate or not, he would mention the invitation, along with his plans to meet Okun, Zeldovich, Dolgoshein, and other renowned Russian physicists.

On the evening before his departure, we had dinner together. Once again, he brought up the story of his prestigious invitation. "Academician Velikhov invited me! I'll have a personal car with a driver!" he boasted.

At the end of supper, I asked Georges, “Would you like to take my mother’s phone number in Moscow?”

“The phone number of your mother? Why?” he asked, surprised.

“Well,” I said, “just in case you run into any trouble in Russia, you could call her.”

“No!” Georges replied confidently. “What trouble could I possibly have? I’m a guest of Academician Velikhov. I’ll have a car…”

Still, Georges was a smart and cautious man. Just before saying goodbye that evening, he muttered, “Well… maybe I’ll write down your mother’s number, just in case. But of course, I won’t actually call her.”

With his uneven handwriting, he jotted the number into his old, worn notebook. The next morning, he left for Moscow, and I quickly forgot all about the small exchange.

At the end of each week, I usually called my mother, and this time I did so on a Thursday evening.

“Ma, how are you?”

“I’m fine,” she replied. “Do you know that Charpak is in my apartment right now?”

I was taken aback. “What is he doing in your apartment?”

“Eating,” my mother said matter-of-factly

A few days later, Charpak returned to his lab at CERN.

“How was your trip?” I asked, genuinely curious.

Georges looked furious. “I couldn’t find a place to eat!” he replied angrily.

It’s worth noting that this took place around 1986–1987 (I don’t recall the exact dates), during a period of severe economic crisis in the USSR. Food stores were nearly empty, even in Moscow, and most restaurants only served foreigners.

Georges then told me what had happened:

"I was desperate to find something to eat, but I couldn't find a decent place anywhere. I went into a restaurant near my hotel. It was completely empty, though the tables were set and waiters were rushing around with plates.

"At the entrance, a waiter stopped me. 'We don't accept anyone,' he said.

"'Why not?' I asked.

"'The restaurant is reserved for foreigners,' he replied.

"'But I *am* a foreigner,' I insisted.

"'No,' the waiter responded with complete indifference. 'We're expecting a group of foreign tourists.'"

"I started pleading with him, I explained I was alone, having trouble finding anywhere to eat, and that letting in just one more person wouldn't make any difference. But the waiter simply replied, 'We have a fixed number of tourists. No.'

I was desperate. I even quietly offered to pay in dollars, but the waiter's expression turned arrogant, and he firmly replied that he didn't accept such currency (VP note: free circulation of U.S. dollars was forbidden in the USSR.).

After three days of similar frustration, I suddenly remembered that you had given me your mother's phone number, and, well, you know what happened next!"

Charpak shoes for one dollar

As often happens with many people, after Georges retired in 1991, many of his friends lost interest in him. He was no longer able to use the technicians or the laboratory facilities at CERN, which made this transition even more difficult. Georges appeared unhappy, and he began to walk with a stooped posture. In what seemed like an instant, a once-celebrated CERN scientist had become a forgotten figure.

During these challenging days, he grew especially close to me and my family. Often, without warning, he would drop by our small apartment in Geneva. My wife, an excellent cook, prepared delicious Russian dishes like pancakes and dumplings, which he thoroughly enjoyed. I vividly recall one of these dinners when Georges, with a theatrical flair, rolled his eyes upward and said, "I believe that one of these beautiful days, the ceiling will collapse and a rich widow will fall into my arms."

After he left, my wife and I exchanged glances and concluded: incorrigible dreamer and optimist. At that moment, his dreams seemed impossible.

Since Georges was no longer a group leader, I lost many of the benefits of being at CERN. So, I decided to accept an invitation for a permanent research position at the Fermi National Accelerator Laboratory (Fermilab) in the USA. This is how, in April 1992, my family and I ended up in America. For a time, I lost contact with Georges.

But life sometimes brings pleasant surprises. In October 1992, Georges was awarded the Nobel Prize in Physics and suddenly became a millionaire, just as he had dreamed! Naturally, I was one of the first to try and congratulate him by phone. His line was always busy, but I eventually got through. I wanted to say something, but instead, we both spontaneously started laughing. That's how we shared our happiness. Then I asked Georges about his immediate plans. "My priority is to buy a good pair of shoes since mine have been leaking for a long time," he said, and we started laughing again.

In the spring of 1993, Georges, at the initiative of Leon Lederman, a former Fermilab director, was invited to visit Fermilab. A few days before Charpak's arrival, Fermilab posted announcements about his visit and the lecture he would give on his work in the main auditorium. I had a dilemma: should I go to the airport to meet him? After all, everything had changed, Georges was no longer just Georges; he was now a Nobel laureate. I shared my thoughts with my wife: "What's the point of going? There will be a crowd at the airport, reporters, Fermilab representatives. I'll be invisible." She replied, "No, you should still go. He was your former boss. Even if he doesn't notice you at the airport, it's no big deal. At least your conscience will be clear. Buy a bottle of champagne, and if you don't need it at the airport, we'll enjoy it ourselves at home." Following this wise advice, I headed to Chicago airport. The plane was scheduled to arrive late in the evening, around 10:00 p.m. To my great surprise, there were no reporters or fellow physicists from Fermilab. The arrival hall was empty, I was the only one there to meet this flight.

Georges was among the first passengers to disembark. He had most likely flown business class. Ever the subtle diplomat, Georges quickly assessed the situation and didn't bat an eyelid. We hugged, shared a sip of champagne, and then I drove him to the Lederman house.

When I arrived at Fermilab the next morning, I noticed that all the announcements for Charpak's lecture, which was supposed to take place the following day, had disappeared. Curious, I asked Dave Andersen, who had previously worked for Charpak at CERN, what had happened. His response was surprisingly rude: "It's none of your business!

A day later, I accidentally ran into Charpak in the lobby of Fermilab. He greeted me with, "Vladimir, after my lecture, tell me about your work here."

"Where and when is your lecture?" I asked cautiously.

"At 10:00 in the main auditorium," he replied casually.

Soon after, Andersen found me and informed me that the lecture had been moved to a meeting room on the second floor, or maybe it was the third, I don't remember exactly.

At five minutes to 10 a.m., I entered the newly appointed location for Charpak's lecture. It was a small room, roughly 20-22 square meters in size. Inside were Charpak, Lederman, seven people from Andersen's group, and a few students.

I sensed that something was off. At 10:00, Lederman introduced Charpak, and Georges, in his usual charming manner, began his lecture. He started by noting that while he was well-known as the inventor of the wire chambers, he had also made significant contributions to the development of spark counters. He then mentioned that it was his idea to measure the third coordinate of particle tracks using the drift time of the primary electrons.

After this introduction, he explained that he wouldn't describe the wire chamber in detail, as Fermilab had its own experts who built and used them in experiments. Instead, he said, "I'll tell you today only about the new developments happening in my group." Charpak then grew excited and began to enthusiastically discuss the optical imaging of particle tracks. He pointed out that developing such innovative detectors would not have been possible without the experience he gained with spark and streamer chambers.

He dedicated the last part of his lecture to the work on photosensitive gaseous detectors combined with solid photocathodes done by his group. He concluded by saying that the Nobel Prize in gaseous detectors would likely go to those who develop large-area proportional chambers with sensitivity to visible light comparable to vacuum photomultipliers.

There wasn't a single question from the audience.

The next day, I invited Georges to our house for dinner. When he arrived, the first thing I asked was if he had bought himself new shoes. "No, I've been so busy these past few days, but I'm planning to do it tomorrow," he replied.

"Can't you sell me your old shoes as a souvenir then?" I joked. He laughed and said, "Okay, that's more profitable than just throwing them away."

A couple of days later, I was pleasantly surprised when Charpak found me in my office and handed me a box containing the shoes.

"How much do I owe you?" I asked.

"One dollar," he replied.

So, I bought Charpak's shoes for one dollar. I kept them for a long time, but eventually, my son, who adored Georges, asked me for them. He even wore the shoes for a while, they became a kind of talisman for him.

When Charpak left Fermilab, I tried to understand the reason for his cold reception there. Why was he met with such indifference? I believe this is a common occurrence in history, where contemporaries often struggle to accept revolutionary ideas. For example, when the Eiffel Tower was built, people were outraged, thinking it spoiled the city's landscape. Now, it's considered the most beautiful landmark in Paris. Similarly, some of Tchaikovsky's works weren't immediately recognized as brilliant by his contemporaries.

Photograph 9. Leon Lederman and Georges Charpak playing table tennis at the Fermilab Recreation Center. This photo was taken by V. Peskov during Charpak's visit to Fermilab in the spring of 1993.

I hope our book will clearly convey the immense contribution Charpak made to the experimental techniques of high-energy physics. After reading it, I believe no one will doubt that Charpak truly deserved the Nobel Prize.

My Last Meeting with Georges in Autumn 2009

My last meeting with George took place in the autumn of 2009 in Saint-Étienne. He had invited me there to demonstrate our flame detector, which we had previously developed together with my colleagues from Saint-Étienne and CERN, to "some important people." Among them were the rector of the École des Mines de Saint-Étienne, the president of the French Academy of Sciences, and the president of the Swedish Academy of Sciences. The demonstration went well: we showcased our detector in full operation and convincingly demonstrated that it was a thousand times more sensitive than any commercially available flame detector at the time.

When the guests left, Georges stayed behind in the lab, and we continued running a few more tests together. One of our team members from Saint-Étienne remarked, “Georges, if we manage to solve the problem of today’s forest fires with this detector, it could mean another Nobel Prize for you!”

Georges promptly replied, “Real scientists never think about awards. They work for the beauty of discovery and science.”

When he eventually left the lab, I walked with him and picked up the conversation we had started earlier. “Tell me, Georges,” I asked, “how did you react when you found out you'd been awarded the Nobel Prize?”

Georges was still riding the high from the successful demonstration and was in excellent spirits that day. He shared with me the following story:

“It was a grey day, just like a hundred others. I had a dentist appointment at 11 a.m., but before heading out, I stopped by my office at CERN to discuss our progress on MICROMEGAS with Giomataris. While we were talking, the phone rang. I picked up the receiver, and an unfamiliar voice asked, ‘Are you Prof. Charpak?’

‘Yes, indeed,’ I replied.

‘I’m a member of the Nobel Prize Committee, and I’m calling to congratulate you, you’ve been awarded the Nobel Prize today!’

Of course, I thought it was a prank and told the caller as much. The person then offered to have someone I knew from the committee call me back and mentioned that a confirmation fax would be sent in the meantime.

After I hung up, I immediately asked Giomataris to leave my office, I needed some privacy. Looking puzzled, he left, and I sat there, trying to make sense of what had just happened. Despite everything, I still believed it might be a joke. I knew that C. Rubbia and a few other physicists had submitted recommendations for me to the Nobel Committee, but I never thought I had a realistic chance, at least not until that day. As I already told you, only bad scientists dream about awards. But just a few minutes later, the phone rang again. This time, it was someone from the committee I knew well, confirming the news. Almost simultaneously, my assistant knocked on the door, dancing with excitement, holding the fax in her hands. At first, Giomataris and the others didn't quite understand what was happening, but once the message sank in, the congratulations started pouring in.

My technician, Bouclier, rushed out to buy a bottle of champagne. Despite everything, I managed to stay composed, even remembered to call my doctor to cancel the dental appointment. That way, I saved some money, otherwise, he would've charged me too much! Don't you agree that was a pretty good reason to cancel?

I always appreciated the contributions of my technicians, and later, I shared part of my award with them. They deserved it."

"And how did you spend the rest of the money?" I asked.

"I bought shoes. I didn't have good shoes before, and on rainy days they leaked."

With those words, Georges stepped into the office of the rector he was scheduled to meet, and I never saw him again. His health declined quickly, he grew weak and lost his mobility. Of course, we stayed in touch by phone. He called me every day, sometimes even twice, to talk about our work on flame and radon detection.

I can say without hesitation: his motivation and driving force truly were the pure beauty of science.

Photograph 10. Georges Charpak with one of the authors of this book (VP.)

Photograph 11. The life of Georges Charpak: a mini-exhibition all about the winner of the 1992 Nobel Prize for Physics in the Globe. 17 Aug 2005, CERN image.

REFERENCES

Aad, G., Abat, E., Abdallah, J., Abdelalim, A.A., et al., 2008, "The ATLAS experiment at the CERN Large Hadron Collider", JINST 3 S08003, 1-407

Aarino, P., Abie, H., Abreu, P., Adam, W., et al., 1991, "The DEPLHI detector at LEP", Nucl. Instrum. and Methods in Phys. Res., A303, 233-276

Abachi, S., Abolins, M., Acharya, B.C., Adam, I., et al., 1994, "The DØ detector", Nucl. Instrum. and Methods in Phys. Res., A338, 185-253

Abbrescia Marcello, Fonte Paulo, Peskov Vladimir, 2018, "Resistive gaseous detectors: designs, performance, and perspectives", Weinheim: Willey-VCH, 1-379

Abbrescia, M., De Cataldo, G., Di Mauro, A., Martinenego, P., et al., 2020, "Evaluation of high sensitivity flat-panel gaseous detectors of flames, sparks and smoke", JINST 15 T12001, 1-23

Abbrscia Marcello, De Cataldo Giacinto, Di Mauro Antonio, Martinengo Paolo, et al., 2020, "Supersensitive multipurpose/multifunctional avalanche gaseous detectors for environmental, hazard, intrusion system (SMART)", arXiv:2011.07297, 1-5

Abe F., Amidei, D., Apollinari, G., Ascoli, G., et al., 1988, "The CDF detector: an overview", Nucl. Instrum. and Methods in Phys. Res., A271, 387-403

Abrams, G., Adolphsen, C.E., Akerlof, C., Alexandes, J.P., et al., 1989, "The Mark II detector for the SLC", Nucl. Instrum. and Methods in Phys. Res., A281, 55-80

Acharrya, B., Alexander, J., Benes, P., Bergmnn, B., et al., 2020, "Timepix3 as a solid-state time-projection chamber in particle and nuclear physics", Proc. of Science, ICHEP2020, 720, 1-8

Adeva, B., Aguilar-Benitez, M., Akbari, H., Alcaraz, J., et al., 1990, "The construction of the L3 experiment", Nucl. Instrum. and Methods in Phys. Res., A289, 35-102

Ajima, Y., Anraku, K., Haga, T., Higashi, Y., et al., 2000, "A superconducting solenoidal spectrometer for a balloon-borne experiment", Nucl. Instrum. and Methods in Phys. Res., A443, 71-100

Albajar, C., Bezaguet, A., Cennini, P., Cittolin, S., et al., 1989, "Studies of intermediate vector boson production and decay in UA1 at the CERN protio-antiproton collider", Z. Phys., C44, 15-61

Alegria Feio, M., Policarpo, A., Charpak, G., Sauli, F., 1980, "Scintillation efficiency of gas proportional scintillation counters", Nucl. Instrum. and Methods, 176, 473-476

G.D. Alkhazov, G.D., 1969, "Main value and variance of gas amplification in proportional counters" Nucl. Instrum. and Methods, 75, 161-162

Alexandreas; d., Bartoli, B., Basteri, D., Biral, R., et al., 1998, "The CLUE experiment operating with 4 telescopes in La Palma", Nucl. Instrum. and Methods in Phys. Res., A409, 488-491

Alexopoulos, T., Altintas, A.A., Alviggi, M., Arik, M., et al., 2009, "The ATLAS muon Micromegas R&D project: towards large-size chamber for the s-LHC", JINST 4 P12015, 1-12

Alexopoulos, T., Burnens, J., Oliveira, R., Glonti, G., et al., 2011, "A spark-resistance bulk-micromegas chamber for high-rate applications", Nucl. Instrum. and Methods in Phys. Res., A640, 110-118

Ahmad N., Ahmad T., Aziz I., Irfan M., et al.,1995, "Technical Proposal for A Large Ion Collided Experiment at the CERN LHC. The ALICE collaboration", 1-237
(https://cds.cern.ch/record/293391/files/cer-000214817.pdf)

Ahmet, K., Akrawy, M.Z., Alexander, G., Allison, J., et al., 1991, "The OPAL detector at LEP", Nucl. Instrum. and Methods in Phys. Res., A305, 275-319

Acharya S. et at. ALICE collaboration. "TRD Technical Design Report", 2004, ALICE-DOC-2004-009 v.1., 88-127a
(https://www.sciencedirect.com/science/article/pii/S0168900217309981?via%3Dihub)

Acharya S., Adam J., Adamová D., Adler C. et al., 2018, "The ALICE Transition Radiation Detector: Construction, operation, and performance", Nucl. Instrum and Methods in Phys. Res., 88-127

Acharay, S., Acosta Hernandz, R., Admova, D., Adler, A., et al., 2024, "ALICE upgrades during the LHC long shutdown 2", JINST 19 P05062, 1-153

Ackermann, K.H., Adams, N., Adler, C., Ahammed, Z., et al., 2003, "STAR detector overview", Nucl. Instrum and Methods in Phys. Res., A499, 624-632

Adeva, B., Barber, D.P., Bckr, U., Berdugo, K., et al., 1984, "A summary of experimental results from Mark J: High energy $e^{+}e$- collisions at PETRA", Phys. Report, 109/3-4, 131-226

Adolfsson, J., Ahmed, M., Aiola, S., Alme, J., et al., 2021, "The upgrade of the ALICE TPC with GEMs and continuous readout", arXiv:2012. 09518, 1-87

Afanasiev, S., Alber, T., Appelshauer, H., Bachler, J., et al., 1999, "The NA49 large acceptance hadron detector", Nucl. Instrum. and Methods in Phys. Res., A430, 210-244

Ahmad, N., Ahmad, T., Aziz, I., Irfan, M., et al., 1995, "Alice technical proposal for a large ion collider Experiment at the CERN LHC", CERN/LHCC/95-71, LHCC/P3, 1-237.

Aihara, H., Alston-Garnjost, M., Badke, D.H., Bakken, J.A., et al., 1983, "Performance of the signal processing system for the time projection chamber", IEEE Trans, Nucl. Sci., NS-30/1, 162-168

Albrecht, H., Binder, U., Bockmann, P., Drews, G., et al., 1989, "Argus: a universal detector at DORIS II", Nucl. Instrum. and Methods in Phys. Res., A275, 1-48

Alexander, J.P., Baggs, R., Fujino, D., Hayes, K., et al., 1989, "The MARK-II vertex drift chamber", Nucl. Instrum. and Methods in Phys. Res., A283, 519-527

Alme, J., Alt, T., Andrey, C., Anguelov, V., et al., 2023," Correction of the baseline fluctuations in the GEM-based ALICE TPC," JINST 18 P11021, 1-28

Almedia, J., Berger, H., Braem, A., Coluzza, C., et al., 1994, "Development of large area fast-RICH prototypes with pad readout and solid photocathodes," Nucl. Instrum. and Methodes in Phys. Res., A348, 216-222

Amaro, F.D., Antonietti, R., Baracchini, E., Benussi, L., et al., 2023 "The CYGNO experiment: a directional Dark Matter detector with optical readout", JINST 18 C09010, 1-4

Anderhub, H., Anders, H., Ansari, S., Boehm, A., et al., 1988, "Operating experience with the MARK J time expansion chamber", Nucl. Instrum. and Methods in Phys. Res., A265, 50-59

Anderson, D.F., Kwan, S., Peskov, V., Hoeneisen, B., 1992," Properties of CsI and CsI-TMAE photocathodes", Nucl. Instrum. and Methods in Phys. Res., A323, 626-634

Anisimov, Yu.S., Cherneneko, S.P., Ivanov, A.B., Kheiker, D.M., et al., 1980, "Multiwire proportional chambers in the device for molecular biology constructed at JINR", Nucl., Instrum. and Methods, 176, 67-69

Angeline, F., Bellazzini, R., Brez, A., Massai, M.M., et al., 1988, "Digital imaging with advanced gaseous detectors developed at Pisa University", Nucl. Instrum. and Method in Phys. Res., A269, 430-435

Antoni, T., Apel, W.D., Badea, F., Bekk, K et al., 2001, "Electron, muon, and hadron lateral distributions measurements in air showers by the KASCADE experiment", Astroparticle Phys., 14, 245-260

Aoyama Takahiko, 1985 "Generalized gas gain formula for proportional counter", Nucl. Instrum, and Methodes in Phys. Res., A234,15-131

Apel, W.D., Arteaaga, J.C., Badea, A.F., Bekk, K., et al., 2010, "The KASCADE-Grande experiment., "Nucl. Instrum. and Methods in Phys. Res., A620, 202-216

Arnold, R., Baillon, P., Besh, H.J., Bosteels, M., et al., 1988, "A ring imaging Cherenkov detector, the DELPHI barrel RICH prototype: Part B: Experimental studies of the detector performance for particle identification". Nucl. Instrum. and Methods in Phys. Res., A270, 289-318

Arnison, G., Astbury, A., Aubert, B., Bacci C. et al., 1983, "Experimental observation of isolated large transverse energy electrons with associated missing energy at s=540 GeV", Phys. Lett. B122, 103-116.

Arnison, G., Astbury, A., Aubert, B., Bacci C. et al., 1983, "Experimental observation of lepton pairs of invariant mass around 95 GeV/c2 at the CERN SPS collider", Phys. Lett. B126, 398-410

Attie David, Aune Stephan, Berthoumieux Eric, Bossu Franceso et al., 2021, "Current status and future developments of Micromegas detectors for physics and applications, "Appl. Sci. 11/5362, 1-35

Aubert, J.J., Becker, U., Biggs, P. J., Burger, J,. et al., 1974, "Experimental Observation of a Heavy Particle J", Phys. Rev. Letters, 33, N 23, 1404-1406.

Augustin, J.-E., Boyarski, A. M., Breidenbach, M., Bulos, F., et al., 1974, "Discovery of a Narrow Resonance in e+ e- Annihilation", Phys. Rev. Letters, 33, N 23, 1406-1408.

Bagnaia, G., Banner, M., Battiston, R., Bloch Ph., et al., 1983, "Evidence for $Z^0 \rightarrow e^+e^-$ at the CERN p collider ", Phys. Lett., B129, 130-140.

Banne, M., Battiston, R., Bloch, Ph., Bonaudi, F. et al., 1983, "Observation of single isolated electrons of high transverse momentum in events with missing transverse energy at the CERN pp collider", Phys. Lett. B122, 476-485.

Barbarito, Elio, Bellotti, R., Cafagna, F., Castellano, M., et al., 1995. "A transition Radiation Detector for positron identification in a balloon-borne particle astrophysics experiment", Nucl. Instrum. and Methods in Phys. Res., A357 588-600

Barile Francesco, 2014, "Light (Hyper-)nuclei production at the LHC measured with ALICE", https://arxiv.org/abs/1411.1941, 1-8

Baru, S.E., Khabakhpashev, A.G., Makarov, I.R., Savinov, G.A., et al., 1985, "Digital X-ray imaging installation for medical diagnostics", Nucl. Instrum. and Methods in Phys. Res., A238, 165-169

Barwick, S.W., Beatty, J.J., Bower, C.R., Chaput, C., et al., 1997, "The high-energy antimatter telescope (HEAT): an instrument for the study of cosmic-ray positrons", Nucl. Instrum. and Methods in Phys. Res., A400, 34-52

Barwick, S.W., Beatty, J.J., Bhattacharyya, A., Bower, C.R., et al.,1997, "Measurements of the cosmic-ray positron fraction from 1 to 50 GeV". Astroph. Journ, 482, L191-L194

Bayatian, G.L., Grigorian, N.K., Khachatrian, V.G., Margarian, A.T., et al., 1994, "Technical proposal. CMS collaboration", CERN/LHCC 94-38, LHCCI/P1, 1-254

Bella, F., Franzinetti, C., Lee, D.W., 1953, "On spark counters", IL Nouvo Cimento, 10#9,1338-1340

Bella, F., Franzinetti, C., 1953, "Spark couters., "IL Nouvo Cimento, 10#10,1461-1479

Belmont, R., 2019," Overview of results from the PHENIX collaboration", Nucl. Phys. A 982, 22-28

Beole, S., Dellacasa, G., Ramello, L., Scalas, E. et al., 1998, "Technical Design Report for The High Momentum Particle Identification detector", CERN/LHC 98-19, 1-188, (https://alice-collaboration.web.cern.ch/sites/default/files/Documents/PROJECTS/HMPID/HMPID_TDR.pdf)

Bettini, Alessandro, 2014, "Introduction to Elementary Particle Physics", Cambridge University Press, second edition. ISBN 978-1-107-05040-2. 1-474

Bhadra, S., Cadabeschi, M., De Perio, P., Galymov, V., et al., 2013, "Optical transition radiation monitor for the T2K experiment", Nucl. Instrum. and Methods in Phys. Res., A703, 45-58

Blum Walter, Riegler Werner, Rolandi Luigi, 2008, "Particle detection with drift chambers" Springer-Verlag Berlin Heidelberg, 157-180

Boezio, M., Barbiellini, G., Bonvicini, V., Schiavon, P., et al., 2001, "Measurements of cosmic-ray electrons and positrons by the Wizard/CAPRICE collaboration", Advances in Space Res., 27, 669-674

Boezio, Mirko, 2005, "The PAMELA Space Experiment," Proceedings of 29th International Cosmic Ray Conference, Pune 10, 255-270.

Bogomolov, G.D., Dubrovske, Yu. V., Peskov, V.D., 1978. "Multiwire Gas Counter for Coordinate Measurements in the VUV Region," Instr. Exp. Tech., 21, 778-783

Bogomolov, G.D., Dubrovske Yu.V., Peskov, V.D., 1978, "Photoelectrical Readout for Multiwire Counters", Instr. Exp. Tech. 21, 639-643

Bogomolov, G.D., Dybrovske, Yu. V., Letunov, A.A., Peskov, V.D., 1987, "Experimental investigations of streamer filaments in high frequency plasma column under high pressure", Soviet Phys. JETF, 93, 519-528

Böhmer, V., 1973, "New Investigations into the Detection Properties of Hybrid Chambers", Nucl. Instrum. and Meth., 107, 157-163

Bohmer, F.V., Ball, M., Dorheim, S, Hoppner, C., et al., 2013, "Simulationof space-charge effect in anungated GEM-based TPC", Nucl. Instrum. and Methods in Phys. Res., A719, 101-108

Borovik-Romanov, A., Peskov, V., 1994, "Cs based photocathodes for gaseous detectors", Nucl. Instrum. and Methods in Phys. Res., A348, 269-274

Bouclier, R., Charpak, G., Dimcovski, Z., Fischer, G., et al., 1970, "Investigation of some properties of multiwire proportional chambers", Nucl. Instrum. and Methods, 88, 149-161

Bouclier, R., Charpak, G., Cattai, A, Million, G., et al., 1983, "Progress in Cherenkov Ring Imaging, Part.1. Detection and localization of photons with multistep proportional chamber", Nucl. Instrum, and Methods, 205, 403-423

Braem, A., Charpak, G., Peskov, V., Suli, F., et al., 1991, "Development of a high-efficiency gaseous detector for the localization of photons from scintillators and new generation of Cherenkov counters", CERN/DRDC/ 92-20, DRDC/P38, 1-15

Braem, A., De Cataldo, G., Davenport, M., Di Mauro, A., et al., 2005, "Results from the aging studies of large CsI photocathodes exposed to ionization radiation in gaseous RICH detector", Nucl. Instrum. and Methods in Phys. Res., 553, 187-195

Braem, A., De Cataldo, G., Davenpot, M., Di Mauro, A., et al., 2007, "The current progress of the ALICE Ring Imaging Cherenkov Detector", Nucl. Instrum. and Methods in Phys. Res., A580, 1144-1147

Breskin, A., Charpak, G., Sauli, F., Atkinson, M., et al., 1975, "Recent observations and measurements with high-accuracy drift chambers", Nucl. Instrum. and Methods 124, 189 -214

Bressani, T., Charpak, G., Rahm, D., Zupancic, C., 1969, "Track localization by means of a drift chamber", Proc. Int. Seminar Filmless spark and streamer chambers (Dubna, April 15-18, 1969), 275-281 (https://cds.cern.ch/record/1221302/files/Bressani.pdf])

Breuil Philippe, Peskov Vladimir, Martinengo Paolo, 2016, "Detector and methods for detecting ultraviolet radiation", US Patent application publication, US 2016/0163492 A1, 1-8

Breyer, B., 1973, "Pulse height distribution in low energy proportional counter measurements", Nucl. Instrum. and Methods, 112, 91-91

Bronilnes Krajcar, Grosswendt Bernd, 2000, "Townsend ionization coefficients of some argon-based mixtures in strong nonuniform electric fields," Journ. Appl. Phys., 48, 6192-6200

Brunbauer Florian, Garcia Francisc, Korkalainen Tero, Alois Lugstein, Alois, et al., 2018, "Combined Optical and Electronic Readout for Event Reconstruction in a GEM-Based TPC", IEEE Trans. Nucl. Sci, 65. No 3 913-918

Buskulis, D., Casper, D., De Bonis, I., Decamp, D., et al., 1995, "Performance of the ALEPH detector at LEP", Nucl. Instrum. and Methods in Phys. Res, A360, 481-506

Byrnes, N.K., Parmaksiz, I., Adams, C., Asaadi, J., et al., 2023 "NEXT-CRAB-0: A High-Pressure Gaseous Xenon Time Projection Chamber with a Direct VUV Camera-Based Readout", arXiv:2304.06091v2 [physics.ins-det], 1-30

Charpak, G, Massonet, L., 1963, "Some developments of the discharge chamber", Rev. Sci. Instr., 34/6, 664-669

Charpak, G., Bouclier, R., Bressani, T., Favier, J., et al., 1968, "The use of multiwire proportional counters to select and localize charged particles" Nucl. Instrum. and Methods, 62, 262-268

Charpak, G., Rahm, D., Steiner, H., 1970, "Some developments in the operation of multiwire proportional chambers", Nucl. Instrum. and Methods, 80, 13-34

Charpak, G., Fischer G., Minten, A., Naumann, L., et al., 1971, "Some features of large multiwire proportional chamber," Nucl. Instrum. and Methodes 97, 377-388

Charpak, G., Sauli, F., 1973, "High-accuracy drift chambers and their use in strong magnetic fields, "Nucl. Instrum. and Methods., 108, 413-426

Charpak, G., Petersen, G., Policarpo, A., Sauli, F., 1978, "Progress in high-accuracy proportional chambers", Nucl. Instrum and Methods, 148, 471-482

Charpak, G., Sauli, F., 1978, "The multistep avalanche chamber: a new high-rate, high-accuracy gaseous detector", Phys. Letter, 78B/4, 523-528

Charpak, G., Melhart, G., Peteresen, G., Sauli, F., 1978, "A New Approach to High-Rate Particle Detectors ", CERN 78-05, EP Division,1-21

Charpak, G., Majewski, S., Melchart, G., Sauli, F., et al., 1979, "Detection of far-ultraviolet photons with the multistep avalanche chamber. Application to Cherenkov light imaging and to some problems in high-energy physics," Nucl. Instrum. and Methodes 164, 419-433

Charpak, G., Melchart, G., Petersen, G., Sauli, F., 1979, "High-accuracy localization of minimum ionizing particles using the cathode-induced charge centre-of-gravity readout," Nucl. Instrum. and Methods, 167, 455-464

Charpak, G., Peisert, A., Sauli, F., Cavestro, A., et al., 1981 "Cherenkov ring imaging using a television digitizer", Nucl. Instrum. and Methods, 180, 387-396

Charpak, G., Dominik, W., Fabre, J.P., Gaudaen, J., et al., 1988, "Some Applications of the Imaging Proportional Chamber", "IEEE Trans. Nucl. Sci., 35, 483-486

Charpak, G., Peskov, V., Scigocki, D., Valbis, J., 1989, "New scintillators for photosensitive gaseous detectors", Proceedings of the Symposium on Particle Identification at High Luminosity Hadron Colliders (eds T. Gourlay and J. Morfin), Fermilab, Batavia, IL, 295-308

Charpak, G., Peskov, V., Sauili, F., Scigocki, D., 1989, "Ethylferrocene in gas, condensed or adsorbed phases: three types of photosensitive elements for use in gaseous detectors", Nucl. Instrum. and Methodes in Phys. Res., A277, 537-546

Charpak, G., Peskov, V., Scigocki, D., et al., 1990, "R&D proposal for LHC crystal calorimeter readout by gaseous (and liquid) detectors with photocathodes", CERN-DRDC/90-77; DRDC/P20, 1-25

Charpak, G., Giomataris, Y., Peskov, V., et al., 1991, "Development of a high-efficiency gaseous detector for the localization of photons from scintillators and Cherenkov radiators" CERN/DRDC 91-35; DRDC/M-11, 1-12

Charpak, G., Fonte, P., Peskov, V., Sauli, F., et al., 1991, "Investigation of operation of a parallel-plate avalanche chamber with a CsI photocathode under high gain condition", Nucl. Instrum. and Methods in Phys. Res., A307, 63-68

Charpak Georges, Giomataris Ioannis, Rebourgeard Philippe, Robert Jean-Pierre, 1995 (priority 11.10.95), "High–resolution position detector for high-flux ionizing particles stream," Patent WO 97/14173, 1-30; (https://patentimages.storage.googleapis.com/ca/41/48/789154ca419cc8/WO1997014173A1.pdf)

Charpak, Georges, 1995, "Research on Particle Imaging Detectors ", World Scientific, 1-650

Charpak, G., Benaben, P., Breuil, P., Peskov, V., 2008, "Detectors for alpha particles and X-rays operating in ambient air in pulse counting mode or/and with gas amplification", JINST 3 P02006, 1-18

Charpak, G., Benaben, P., Breuil, P., Martinengo, P., et al., 2008, "Development of new hole-type avalanche detectors and the first results of their applications," IEEE Trans. Nucl. Sci., 55, 1657-1663

Charpk, G., Benaben, P., Breul, P, Martinengo, P., et al., 2009, "Progress in the development of a SRETGEM-bases detector for an early forest fire warning system" JINST 4 P1 2007, 1-16

Charpak, G., Benaben, P., Breuil, P., Martinengo, P., et al., 2011, "Results from prototypes of environmental and health alarm devices based on gaseous detectors operating in air in counting mode," Nucl. Instrum. and Methods in Phys. Res., A628, 187-189

Chefdeville, M., Oliveira, R., Drancourt, C., Geffroy, N., et al., 2021, "Development of Micromegas detectors with resistive anode pads", Nucl. Instrum. and Methods in Phys. Res., A1003/165268, 1-12

Commichau, V., Deutschmann, M., Draheim, K.J., Fritze, P, et al., 1985, "Test of a high resolution drift chamber prototype", Nucl. Instrum. and Methods in Phys. Res., A235, 267-278

Cools, A., Aune, S., Beau, F., Benali, M., et al., 2023, "Neutron and beta imaging with Micromegas detectors with optical readout", Nucl. Instrum. and Methods in Phys. Res., A 1048/167910, 1-6

Cortese P., Dellacasa G., Ramello L., Scalas E., et al., 2001, "TRD Technical Design Report," CERN /LHCC 2001-021 ALICE TDR 9, 1-268 (https://cds.cern.ch/record/519145/files/cer-2275567.pdf)

Curtiss, L.F., 1933, "Simplified automatic Wilson chamber", 1932, Bureau of Standards Journal of Research, 8, 579-582, (http://nvlpubs.nist.gov/nistpubs/jres/8/jresv8n5p579_A2b.pdf)

Dangendorf, V., Breskin, A., Chechik, R., Schmidt-Boecking, H., 1990, "A gas filled UV photon detectors with CsI photocathode for detection of Xe Light", Nucl. Instrum. and Methods in Phys. Res., A389, 322-324

Decamp, D., Deschizeaux, B., Lees, J.P., Minard, M.N., et al., 1990, "ALEPH: A detector for electron-positron annihilations at LEP", Nucl. Instrum. and Methods in Phys. Res., A283, 582-589

Deisting, A., Waldron, A.V., Atkin, E., Barker, G., et al., 2021, "Commissioning of a High Pressure Time Projection Chamber with Optical Readout", Instruments, 5/22, 1-39 (https://doi.org/10.3390/instruments5020022)

Deisting, A., 2022, "Commissioning of a hybrid readout TPC test set-up and gas gain simulations, "Journal of Physics: Conference Series, 2374 /012145, 1-4

Dellacasa G., Ramello L., Scalas E., Sitta, M., et al., 2000, "Technical Design Report of the Time Projection Chamber", 1-226 (https://cds.cern.ch/record/451098/files/open-2000-183.pdf)

Del Sordo Stefano, Abbene Leonardo, Caroli Esio, Mancii Anna Maria, et al., 2009, "Progress in the Development of CdTe and CdZnTe Semiconductor Radiation Detectors for Astrophysical and Medical Applications", Sensors, 9, 3491-3526

Diatroptov, D.B., Peskov, V.D., 1971, "VUV Spectra of Free Plasma Column in UHF Field at High Pressure", Journ. Exper. and Theor. Phys., 61, 1038-1043 (in Russian)

Di Mauro, A., Ljubicic, S., Nappi, E., Paic, G., et al., 1994, "Particle identification with a solid photocathode RICH in ALICE at LHC", Nucl. Instrum. and Methodes in Phys. Res., A343, 284-287

Dimcovski, Z., Favier, J., Charpak, G., Amato, G., 1971, "High-energy charged particles separation by means of a cascade of multiwire proportional chambers, "Nucl. Instrim. and Methods, 94, 151-155

Derre, J., Giomataris, I., Zaccone, H., Bay, A., et al., 2001, "Spatial resolution in Micromegas detector", Nucl. Instrum. and Methods in Phys. Res., A459, 523-531

Dominik, W., Zaganidis, N., Astier, P., Charpak, G., et al., 1989, "A gaseous detector for high-accuracy autoradiography of radioactive compounds with optical readout of avalanche positions", Nucl. Instr. Meth. Phys. Res., A, 278, 779–787

Dorion, I., Ruscev, M., Lilot, A.P., 1987, "A novel unidimensional position sensitive multiwire detector", IEEE Trans. Nucl., Sci., NS-34, 442-448

Feller, J.M., Barbaro-Galtieri, A., Dorfan, J.M., Ely, R., et al., 1977, "The lead-glass wall addition to the SPEAR MARK I magnetic detector", SLAC-PUB-2060, LBL-6466, 1–5.

Fischer, J., Shibata, S., 1972, "The Hybrid Chamber: A Proportional Chamber with Gated Spark Readout", Nucl. Instrum. and Meth., 101, 401-404

Fitch, J., Ling, T.Y., Romanowski, T.A., Smith, E., et al., "A novel proportional drift tube system", Nucl. Instrum. and Methods in Phys. Res., 226, 373-375

Fonte, P., Peskov, V., Sauli, F., 1991, "Feedback and breakdown in parallel-plate chambers", Nucl. Instrum and Methods in Phys. Res., A305, 91-110

Fonte, P., Peskov, V., Sauli, F., 1991, "VUV emission and breakdown in parallel-plate chambers", Nucl. Instrum. and Methods in Phys. Res., A310, 140-145

Fonte, P., Peskov, V., Ramsey, B., 1999, "The fundamental limitations of high-rate gaseous detectors", IEEE Trans. Nucl. Sci., 46, 321-325

Fonte, P., Carolino, N., Costa, L., Ferreira-Marques, R., et al., 1999, "A spark-protected high-rate detector", Nucl. Instrum. and Methods in Phys. Res., A431, 154-159

Fraga, M.M F.R., Fraga F.A.F., Fetal, S.T. G., Margato, L.M.S., et al., 2003, "The GEM scintillation in He-CF_4, Ar-CF_4, Ar-TEA and Xe-TAE mixtures", Nucl. Instrum. and Methods in Phys. Res., A 504, 88-92

Francke Tom, Peskov Vladimir, 2014, "Innovative applications and developments of micro-pattern gaseous detectors," Hershey, PA: Engineering Science Reference, 1-303

Francke, T., Peskov, V., 2016, "Position-Sensitive Gaseous Photomultipliers: Research and Applications", Hershey, PA: Engineering Science Reference, 1 - 562

Gaertner, H.R., Yeater, M.L., 1949, "A fast recycling cloud chamber and pulsed magnetic field equipment for use in pulsed accelerators", Rev., Sci. Instr., 20, 588-595

Gaisser, T. K., Schaffler, R.K. and Robert, K., 1992, "Cosmic-Ray Secondary Antiprotons: A Closer Look" Astrophys. Journal, 394, 174-183

Garibian, G.M., 1958, "Contribution to the theory of transition radiation" Sov. Phys JETF, 1079-1085 (1957, Zh. Exp. Teor. Fiz. 33, 1043-1049)

Garibian, G.M., 1960, "Transition radiation effects in particle energy losses", Sov. Phys. JETF, 37/10, 372-376 (Zh. Exp. Teor. Fiz. 37, 1959, 527-531)

Ginzburg, V.L., Frank I.M., 1945, "Radiation of a uniformly moving electron due to its transition from one medium into another", Journ. Phys. 9, 353-362 (JETP, USSR, 1946 16, 15-28)

Giomataris, I., De Oliveira, R., Andriamonje, S., Aune, S., et al., 2006, "Micromegas in a bulk", Nucl., Instrum. and Methods in Phys. Res., A 560, 405-408

Glashow, S.L. Iliopoulos, J. Maiani, L., 1970, "Weak interactions with lepton–hadron symmetry". Physical Review D. 2 (7), 1285-1292

Goldsmith, P., Jelley, J.V., 1959, "Optical transition radiation from protons entering metal surface, "Philos. Mag., 4, 836-844

Grunsfeld, J.M., 1988, "The cosmic ray nucleus experiment on the Spacelab –II mission" in the book "Genesis and propagation of cosmic rays"; editors Shapir M.M and WefelJ.P NATO ASI series, D. Reidel Publ. Company, 381-390

Guth, A.H., 1981, "Inflationary universe: A possible solution to the horizon and flatness problems," Phys. Rev., D23, 347-356.

Hall, C.J., Lewis, R.A., Parker, B., Sheldon, J., et al., 1995, "Microgap Multiwire Chamber", Nucl. Phys. B (Proc. Suppl.), 44, 242-245

Hoedlmoser, H., Braem, A., De Cataldo, G., Davenport, M., et al., 2006, "Production technique and quality evaluation of CsI photocathodes for the ALICE/HMPD detector", Nucl. Instrum. and Methodes in Phys. Res., A566, 338-350

Hoedlmoser H., Braem A., De Cataldo G., Davenport M., et al., 2007, "Long-term performance and ageing of CsI photocathodes for the ALICE/HMPID detector", Nucl., Instrum. and Methods in Phys. Res., A574. 28-38

Iacobaeus, C., Danielsson, M., Fonte, P., Francke, T., et al., 2002, "Sporadic electron jets from cathodes the main breakdown- triggering mechanism in gaseous detectors", IEEE Trans. Nucl. Sci., 49, 1622-1628

Ivaniouchenkov, I., Fonte, P., Peskov, V., Ferriera-Marques, R., et al., 1998, "The high-rate behaviour of parallel mesh chambers", IEEE Trans. Nucl. Sci., 45, 258-262

Ivaniouchenkov, I., Fonte, P., Peskov, V., Ramsey, B.D., 1999, "Breakdown limit studies in high-rate gaseous detectors", IEEE Trans. Nucl. Sci., 45, 258-262

Jeawons, A.P., Charpak, G., Stubb, R.J., 1975, "The high-density multiwire drift chamber", Nucl. Instrum. and Methods, 124, 491-503

Kahn, R., Fourme, R., Caudron, B., Bosshard, R et al., 1980, "A fast X-ray diffractometer based on a spherical drift multiwire chamber", Nucl. Instrum. and Methods, 172, 337-344

Karabadzak, G.F., Peskov, V.D., Podolyak, E.R., 1983, "Coordinate gas counter with superhigh energy resolution for plasma diagnostics", Nucl., Instrum. and Methods, 217, 56-64

Kataoka, Y., Leontsinis, S., Ntekas, K., 2014, "Performance Studies of a Micromegas Chamber in the ATLAS Environment," arXiv:1310. 8603v2 [physics.ins-det], 1-9

Keuffel, J. Warren, 1949, "Parallel-plate counters", Rev. Sci. Instrum., 20, 202-208

Kosarev, E.L., Peskov, V.D., Podolyak, E.R., 1983, "High resolution soft X-ray spectrum reconstruction by MWPC attenuation measurements", Nucl. Instrum, and Methods, 208, 637-645

Kotriakhova, Sofia, 2020, "Long-term and efficient of the MWPC muon detector at LHCb", Nucl. Instrum. and Methods in Phys. Res., A958, 162787, 1-4

Lacy, J.L., Verani, M.S., Ball, M.E., Roberts, R, 1988, "Clinical applications of a pressurized xenon wire chamber gamma camera utilizing the short lived gent ^{178}Ta", Nucl. Instrum. and Methods in Phys. Res., A269, 369-376

Leith, David W.G.S., 1988, "Status of Cherenkov ring imaging systems", Nucl. Instrum. and Methods in Phys. Res., A265, 120-136

Lipman, C., 2012, "Performance of the ALICE Time Projection Chamber", Physics Procedia 37, 434 – 441

Lowe Adam, Majumdar Krishanu, Mavrokoridis Konstantinos, Philippou Barney, et al., 2020, "Optical Readout of the ARIADNE LArTPC Using aTimepix3-Based Camera," Instruments, 4/35, 1-19

Madaras Ronald, J., Oddone Piermaria, J., 1984, "The time projection chambers", Physics Today 37/8 36-47

Mangeot, P., Coutrakon, G., Hubbard, J.R., Mullie, J., et al., 1983, "Progress in Cherenkov Ring Imaging. Part 2: Identification of charged hadrons at 200 GeV/c", Nucl. Instrum. and Methods, 216, 79-91

Marx Jay N., Nygren David R., 1978, "The Time Projection Chamber", Physics Today, 31(10), 46–53

Maurin, D., Ahlers, M., Dembinski, H., Haungs, A., et al., 2023, "A cosmic-ray database update: CRDB v4.1," arXiv:2306.08901., 1-20.

Miernik, K., Dominik, W., Czyrkowski, H., Dabrowski, R., et al., 2007, "Optical time projection chamber for imaging nuclear decays", Nucl. Instr. Meth. Phys. Res., A, 581, 194–197

Moiseev, A., Yoshimura, K., Ueda, I., Anraku, K., et al., "Cosmic-ray antiproton flux in the energy range from 200 to 600 MeV," Astrophys. Journ. 474, 479-490

Montag Benjamin W., Bellinger Steven L., Edwards Nathaniel S., Lage Jackson, et al., 2019," Recent progress in the commercialization of the Li Foil multi-wire proportional counter neutron detector", Rad. Phys. and Chemistry, 155, 158-163

Nappi Eugenio and Seguinot Jacques, 2005, "Ring Imaging Cherenkov Detectors: The state of the art and perspectives", La Rivista del Nuovo Cimento, 2005, 28, N 8-9, 1-130

Nappi Eugenio, Peskov Vladimir, 2013, "Imaging gaseous detectors and their applications", Wiley-VCH, 1-323

Navas, S., Amsler, C., Gutsche T., Hanhart, C., et al., 2025, "Review of particle physics", Phys. Rev. D 110, 1-2382

Nowak Gregor, Stefanescu Irina, Beldowski Andres, Fenske Jochen, et al., 2023, "A design study of a 1-m^2 multiwire-proportional-chamber positron-sensitive neutron –detector (MWPC-PSND)", EPJ Web Conf., 286, 1-12

Oed, A., 1988, "Position-sensitive detectors with microstrip anode for electron multiplication with gases," Nucl. Instrum. and Methods in Phys. Res., A263, 351-359

Oliveira, R., Peskov, V., Pietropaolo, F., Picchi, P., 2010, "First tests of MICROMEGAS and GEM-like detectors made of a resistive mesh", IEEE Trans. Nucl. Sci., 57, 3744-3752

Pachmayer, Yvonne, 2013, "Physics with the ALICE transition radiation detector", Nucl. Instrum. and Methods in Phys. Res., A706, 6-11

Patra Rajendra Nath, Singaraju, R.N., Biswas, S., Nayak, T.K., et al., 2018, "Time resolution and characteristics study of MWPC detectors with different Ar based mixtures," Springer. Proc. Phys., 203, 509-512

Periale, L., Peskov, V., Carlson, P., Francke, T., et al., 2002, "Detection of the primary scintillation light from dense Ar, Kr, and Xe with novel photosensitive gaseous detectors", Nucl. Instrun. and Methods in Phys. Res., A478, 377–383

Peskov, V.D., Sakharov, Ya. A., Tichenko, E.A., 1973, "Spatially Non-Uniform Filters for Ultrasoft X-Rays", Letters to Journ. Exp. and Theor. Phys., 17, 197-202 (in Russian).

Peskov, V., 1975, "Current oscillations in a positive corona discharge", Sov. Phys. Tech. Phys., 20, 1588-1590

Peskov, V., 1976, "Secondary Processes on Gas Counters I", Sov. Phys. Techn. Phys., 20(6), 791-794

Peskov, V.D., 1979, "VUV Spectrum of Townsend Avalanches", Zh. Prikl. Spectr. 30, 860-865 (in Russian)

Peskov, V.D., 1979, "On the VUV Radiation of Townsend Avalanches", J. Phys. (France) 40, C7-331-C7334

Peskov, V.D., 1980, "Coordinate Gas Scintillation Counters of VUV and X-Ray Photons", Instr. Exp. Tech., 23, 507-512

Peskov Vladimir. 1980, "Complex/Multiparameter Plasma Diagnostics in UV and X-ray Regions." (Doctor of Science/Habilitation Thesis). Kapitza Institute for Physical Problems, Moscow, Russia; file:///C:/Users/Win7/Downloads/1.1-%20(1).pdf)

Peskov, V., Charpak, G., Mine, P., Sauli, F., et al., 1988, "Liquid and solid organic photocathodes", Nucl. Instrum. and Methods in Phys. Res., A269, 149-160

Peskov, V., Charpak, G., Dominik W., Sauli, F., 1989, "Investigation of light emission from a parallel-plate avalanche chamber filled with noble gases and with TEA, TMAE and H_2O vapours at atmospheric pressure", Nucl. Instrum. and Methods in Phys. Res., A277, 547-556

Peskov, V., Borovik-Romanov, A., Sokolova, T., Silin, E., 1994, "Gaseous detectors of ultraviolet and visible photons", Nucl. Instrum. and Methods in Phys. Res. A348, 269-274

Peskov, V., 1995, "Gaseous detectors with solid photocathodes", Proceedings of the 1st International. Four Seas Conference, Trieste, Italy, 1995, CERN 97-06, 48-51 (Retrieved from: http://cds.cern.ch/record/282759/files/ CERN-97-06.pdf)

Peskov, V., Ramsey B.D., Fonte, P., 1997, "Surface streamer breakdown mechanisms in micostrip gas counters", Nucl. Instrum. and Methodes in Phys. Res., A392, 89-93

Peskov, V., Ramsey, B.D., Kolodziejczak., Fonte, P., 1997, "Feedback and breakdowns in microstrip gas counters, "Nucl. Instrum. and Methods in Phys. Res., A397, 243-260

Peskov V., Ramsey, B.D., Fonte, P., 1998, "Breakdown features of various microstrip-type gas counters designs and their improvements", IEEE Trans. Nucl. Sci., 45, 1998, 244-248

Peskov, V., Fonte, P., Danielsson, M., Iacobaeus, C., et al., 2001, "The study and optimization of new micropattern gaseous detectors for high-rate applications", IEEE Trans. Nucl. Sci., 48, 1070-1074

Peskov, V., Fonte, P., 2009, "Research on Discharges in Micropattern and Small Gap Gaseous Detectors", arXiv:0911.0463, 1-44 (https://arxiv.org/abs/0911.0463)

Peskov Vladimir, Di Mauro Antonio, De Oliveira Rui, Breuil Philippe, 2014, "A wall-less electron multiplier," Patent WO 2014/146673 A1, 1-57

Petrvari Judit, Osman Magda, Bhattacharya Joydeep, 2016, "The role of intuition in the generation and evaluation stages of creativity", Frontiers on Psyholog., 7/1420, 1-12

Piuz Francois, 2003, "Ring Imaging CHerenkov systems based on gaseous photo-detectors: trends and limits around particle accelerators" Nucl. Instrum. and Methods in Phys. Res. A, 2003, Vol. 502. 76-90

Protheroe, R., J., 1982, "On the nature of the cosmic ray positron spectrum", Astroph. Journ., 254, 391-397

Raether, H., 1964, "Electron avalanches and breakdown in gases", Butterworths, 1-191

Revol Jean Pierre, 2023. The W and Z at UA1. Symposium. Electroweak milestones - 50 years of neutral currents, 40 years of W and Z bosons. 1-33, Accessed Feb. 2025 (from https://indico.cern.ch/event/1301000/contributions/5598128/attachments/2742458/4772494/WZ.40.years.11.pptx)

Riodran, M., 1992, "The discovery of quarks," SLAC-PUB-5724, T/E, 1-36 (https://s3.cern.ch/inspire-prod-files-9/90ba9674ad34bb45f161327cb8cbf442#:~:text=Instead%2C%20they%20appeared%20to%20be,of%20an%20electron%20or%20proton).

Rutherford, E., Geiger, H., 1908, "An electrical method of counting the number α-particles from radio-active substances", Proc. R. Soc. London. Ser. A81 (546), 141-161

Sauli, F., 1997, "A new concept for electron amplification in gas detectors", Nucl. Instrum. and Methods in Phys. Res., A386, 531-534

Sauli, F., 2014, "Gaseous radiation detectors, fundamentals and applications", Cambridge Univ. Press, 1-497

Seguinot J., Ypsilantis T., 1977, "Photo-ionization and Cherenkov Ring Imaging," Nucl. Instrum. Methods, 142, 377-391

Seguinot, J., Tocqueville, J., Ypsilntis, T., 1980, "Imaging Cherenkov detector: photo-ionization of Triethylamine", Nucl., Instrum. and Methods, 173, 283-298

Seguinot, J., Charpak, G., Giomtaris, I., Peskov, V., et al., 1990, "Reflective UV photocathodes with gas-phase electron extraction: solid, liquid and absorbed thin films", Nucl. Instrum. and Methods in Phys. Res., A297, 133-147

Sharma, A., Sauli, F., 1992, "A measurement of the first Townsend coefficient in Argon-based mixtures at high fields". Nucl. Instrum. and Methods in Phys. Res., A323 280-283

Simon, A., 2020, "The NEXT experiment for neutrinoless double beta decay searches" , Journal of Physics: Conference series 1342/012046, 1-6

Skovoroda, A.A., Zvonkov, A.V., 2001, "A stationary filamentary microwave discharge at a high pressure of argon", Journ. Experim. and Theor. Phys. 92, 78-85

Suzuki, M., Breskin, A., Charpak, G., Daubie, E., et al., 1988, "On optical readout of gas avalanche chamber and its application," Nucl. Instrum. and Methods in Phys. Res., A263, 237-242

Takasaki Fumihiko, 1988, "Experimental results from Tristan", Nucl. Phys. B, Proc. Suppl., 3, 17-37

Titt, U., Breskin, A., Chechik, R., Dangendorf, V., et al., 1998, "A Time Projection Chamber with optical readout for charged particle track structure imaging", Nucl. Instr. Meth. Phys. Res., A, 416, 85–99

Townsend John, 1901, "The Conductivity produced in gases the motion of negatively charged ions", The London, Edinburgh, and Dublin Philosophical Magazine and Journal of Science, "XVII, 1:2, 198-227

Townsend, D., Frey P., Jeavons, A., Reich G., et al., 1987, "High density avalanche chamber (HIDAC) positron camera," J. Nucl. Med, 28 1554-1562

Trumper, J., 1960, "Parallel plattenzahler mit electronisch gesteuerter funkenzundung fur Teilchenbahnbestimmungen", Atomkernenergie, 5, 121-128

Uhlig, K., 1991, "Surface Discharges in High Magnetic Fields", Hartung-Gorre Verlag, 1-220

Volpe, G., Peskov, V., 2018, "A GEM-based detector for detection and imaging of sparks and flames," Il Nuovo Cimento, 41 C, 96, 1-3

Walenta, A.H., Heintze, J., Schurlein, B., 1971, "The multiwire drift chamber a new type of proportional wire chamber", Nucl. Instrum. and Methods, 92, 373-380

Weinberg, S., 1979, "Cosmological Production of Baryons", Phys. Rev. Lett., 42, 850-853

Yoshimura, K., Orito, S., Anraku, K., Golden, R., et al., 1995, "Observation of cosmic-antiprotons at energies below 500 MeV", Phys. Rev. Lett, 75, 3792-3796

Young, C.C., Atwood, W.B., Carr, J., Chadwick, G., et al., 1986, "Performance of the SLD central drift chamber prototype", IEEE Trans. Nucl. Sci., 33, 176-77

Further readings

For those interested in learning more about MWPCs and other modern gaseous detectors, we recommend the following recent books:

Nappi, Eugenio, Peskov, Vladimir, 2013, "Imaging gaseous detectors and their applications", Wiley-VCH, 1-323

Sauli, Fabio, 2014, "Gaseous radiation detectors. Fundamental and applications", Cambridge. Univ. Press, 1-497

Francke Tom, Peskov, Vladimir, 2014 "Innovative applications and developments of micro-pattern gaseous detectors", Engineering Science Reference, 1-303

Francke Tom, Peskov, Vladimir, 2016, "Position-sensitive gaseous photomultipliers: research and applications", Engineering Science Reference, 1-562

Abbrescia, Marcello, Peskov Vladimir, Fonte Paulo, 2018, "Resistive gaseous detectors, designs, performance and perspectives", Wiley-VCH, 1-379

Sauili, Fabio, 2021, "Micro-pattern gaseous detectors", World Scientific,1-349

ACKNOWLEDGMENTS

We gratefully acknowledge the contributions of Jean-Pierre Revol to the overall structure of the book, Christian Sonnabend for his insightful suggestions on applying AI to the analysis of Time Projection Chamber physics data, and Erlend Flatland for his work on AI applications in HMPID Cherenkov patterns recognition.

We are especially grateful to Gianluca De Cataldo for his careful and thoughtful reading of the draft, offering invaluable feedback from the fresh and insightful perspective of a young reader.

Our sincere appreciation goes as well to the ALICE secretary, Mrs. Nathalie Gouriou, for her kind and efficient assistance in preparing the contract with the publisher.

LIST OF ABBREVIATIONS

ADCs -analog to digital converter
AGS -Alternating Gradient Synchrotron at Brookhaven
AI-Artificial Intelligence
ALEPH-Apparatus for LEP Physics
ALICE-A Large Ion Collider Experiment
AMS-Alpha Magnetic Spectrometer
ATLAS-A Toroidal LHC Apparatus)
BRAHMS-Broad Range Hadron Magnetic Spectrometers
CD-Central Detector
CDF-Collider Detector at Fermilab
CCD-charge-coupled device
CERN-European Particle Physics Laboratory
CERN PS –CERN proton synhrotron
CLUE-Light Ultraviolet Experiment
CMOS -Complementary Metal Oxide Semiconductor
CMS-Compact Muon Solenoid
COMPASS-Common Muon Proton Apparatus for Structure and Spectroscopy
CP violation-violation of the symmetry between matter and antimatte
CPU-central processing unit
CRID-Cherenkov Ring Imaging Detector
CSA-charge-sensitive amplifiers
CT – computer tomography
DC-drift chamber
DELPHI-Detector with Lepton, Photon, and Hadron Identification
DESY-Deutsches Elektronen-Synchrotron
DØ –Detector at Zero Degree at Fermilb
E-electric field
EAS-extensive air showers
EF-Ethylferrocene
EM - the electromagnetic calorimeter
FCC-Future Circular Collider

FEE-Front-End Electronics
Fermilab – Fermi National Accelerator Laboratory in the USA
FWHM-full width at half maximum
GEM –gas electron multiplier
GIM mechanism -Glashow–Iliopoulos–Maiani, mechanism
GPUs-Graphics Processing Units
GSI-Gesellschaft für Schwerionenforschung (Society for Heavy Ion Research)
HADES- High Acceptance DiElectron Spectrometer
HERA-Hadron-Electron Ring Accelerator
HMPID –high momentum particle identification
HT-Hough Transform algorithm,
HTM-Hough Transform Method
HV –high voltage
INFN - Istituto Nazionale di Fisica Nucleare
IR-infrared
KASCADE-KArlsruhe Shower Core and Array Detector
KEK-High Energy Accelerator Research Organization in Japan
LEP-Large Electron-Positron Collider,
LHC-Large Hadron Collider
L3-
MICROMEGAS
MCP-Microchannel Plate
ML-machine learning
MIP-minimum ionizing particles
MSGC-microstrip gas counter
MWPC- multiwire proportional chamber
NASA-National Aeronautics and Space Administration
NC-neutral current
OPAL-Omni-Purpose Apparatus for LEP
PEP-positron-electron-proton
PETRA-electron-positron colliding beam accelerator
PICO experiment-Project in Innovative COld technology
PID- Particles identification
PIND-photionization detector
PMT-photomultipliers tube

PCB –printed circuit board
PEP-Positron Electron Project
PET -Positron Emission Tomography
PHENIX-Pioneering High Energy Nuclear Interaction Experiment
PPAC-parallel- plate avalanche chamber
P_t-transverse momentum
Q - quantim efficiency
QCD-Quantum Chromodynamics
QGP-Quark–Gluon Plasma
R&D -research and development
RHIC-Relativistic Heavy Ion Collider
RICH-Ring Imaging Cherenkov counters
SLAC-Stanford Linear Accelerator Center
SLD -SLAC Large Detector
SLC- Stanford Linear Collaider
SNR- signal to noise ratio
SPEAR-Positron-Electron Asymmetric Ring
SPS-Super Proton Synchrotron
STAR-Solenoidal Tracker at RHIC
TDC-Time-to-Digital Converter
TEA-Triethylamine
THGEM-thick gas electron multiplie
TMA-Trimethylamine
TMAE-Tetrakis(dimethylamino)ethylene
TMS-tetramethylsilane
TOF-Time-of-Flight
TPC- Time Projection Chamber
TR-transition radiation
TRD-Transition Radiation Detector
UHF-ultrahigh-frequency microwaves
UV – ultraviolet